Lecture Notes in Economics and Mathematical Systems 529

Springer-Verlag Berlin Heidelberg GmbH

Werner Krabs
Stefan Wolfgang Pickl

Analysis, Controllability and Optimization of Time-Discrete Systems and Dynamical Games

 Springer

Authors

Prof. Dr. Werner Krabs
Department of Mathematics
Technical University Darmstadt
Schlossgartenstrasse 7
64289 Darmstadt
Germany

Dr. Stefan Wolfgang Pickl
Department of Mathematics
Center of Applied Computer Science
ZAIK
University of Cologne
Weyertal 80
50931 Cologne
Germany

Cataloging-in-Publication Data applied for

A catalog record for this book is available from the Library of Congress.

Bibliographic information published by Die Deutsche Bibliothek
Die Deutsche Bibliothek lists this publication in the Deutsche Nationalbibliografie;
detailed bibliographic data is available in the Internet at http://dnb.ddb.de

ISSN 0075-8450
ISBN 978-3-540-40327-2   ISBN 978-3-642-18973-9 (eBook)

DOI 10.1007/978-3-642-18973-9

http://www.springer.de

© Springer-Verlag Berlin Heidelberg 2003
Originally published by Springer-Verlag Berlin Heidelberg New York in 2003

Typesetting: Camera ready by author
Cover design: *Erich Kirchner*, Heidelberg

Printed on acid-free paper   55/3143/du   5 4 3 2 1 0

Dedicated to the people of Navrongo

Preface

J.P. La Salle has developed in [20] a stability theory for systems of difference equations (see also [8]) which we introduce in the first chapter within the framework of metric spaces. The stability theory for such systems can also be found in [13] in a slightly modified form. We start with autonomous systems in the first section of chapter 1. After theoretical preparations we examine the localization of limit sets with the aid of Lyapunov Functions. Applying these Lyapunov Functions we can develop a stability theory for autonomous systems.

If we linearize a non-linear system at a fixed point we are able to develop a stability theory for fixed points which makes use of the Fréchet derivative at the fixed point.

The next subsection deals with general linear systems for which we introduce a new concept of stability and asymptotic stability that we adopt from [18]. Applications to various fields illustrate these results. We start with the classical predator-prey-model as being developed and investigated by Volterra which is based on a 2×2-system of first order differential equations for the densities of the prey and predator population, respectively. This model has also been investigated in [13] with respect to stability of its equilibrium via a Lyapunov function. Here we consider the discrete version of the model. If we discretize the original model of interacting growth of populations in terms of two first order differential equations we obtain a general discrete model for interacting logistic growth of two populations. As a last example we present an emission reduction model for the reduction of carbon dioxide emissions.

The next section of chapter 1 deals with non-autonomous systems. Definitions and elementary properties of such are presented. Again, we present the stability theory based on Lyapunov's method for such systems. We regard general systems and as a special case the linear case. As an application we describe the temporal development of the concentration of some poison like urea in the body of a person suffering from a renal disease and having to be attached to an artificial kidney.

The second chapter deals with time-discrete controlled systems. Here we begin with the autonomous case and introduce the problem of fixed point controllability. If we regard linear systems, we obtain the problem of null-controllability. For this we present an algorithmic method for its solution.

Furthermore, we describe the problem of stabilization of controlled systems. Then several applications are presented. We pick up the emission reduction model and concentrate ourself on the controlled system which we linearize at a fixed point. As a second example we treat the controlled prey-predation model. Also a planar pendulum with moving suspension point can be described by that modeling. We consider a non-linear pendulum of length $l(> 0)$ whose moment is controlled by moving its suspension point with acceleration $u = u(t)$ along a horizontal straight line.

In the next section of chapter 2 we regard the non-autonomous case and the specific problem of fixed point controllability. Furthermore, the general problem of controllability, the stabilization of controlled systems and the problem of reachability is treated.

The third chapter deals with the controllability of dynamical games. These are formulated as controlled autonomous dynamical systems which as uncontrolled systems admit fixed points. The problem of controllability consists of finding control functions such that a fixed point of the uncontrolled system is reached in finitely many time steps. For this problem a game theoretical solution is given in terms of Pareto optima in the cooperative case and Nash equilibria in the non-cooperative case.

For the emission reduction model the non-cooperative treatment of the control problem leads to the application of linear programming for the calculation of Nash equilibria and the cooperative treatment gives rise to the application of cooperative game theory.

In particular we have to investigate the question under which conditions the core of such a game is nonempty.

In this connection we also consider n-person goal-cost-games and present a dynamical method for finding a Nash equilibrium in such a game.

After the treatment of evolution matrix games we come back to n-person goal-cost-games which we transfer into cooperative n-person games. Here we investigate the question under which conditions the grand coalition is stable.

The appendix supplies the reader with additional information.
Sections $A.1$ and $A.2$ are concerned with the core of a general cooperative n-person game and a linear production game, respectively. In Section $A.3$ necessary and sufficient conditions for weak Pareto optima of non-cooperative n-person games are given and Section $A.4$ deals with duality in such games. The book ends with bibliographical remarks.

The authors want to thank Silja Meyer-Nieberg for carefully reading the manuscript and Goran Mihelcic for excellent typesetting. He solved every tex-problem which occured in minimal time.

Cologne,   *Werner Krabs*
May 2003   *Stefan Pickl*

Contents

1

Uncontrolled Systems

1.1 The Autonomous Case

1.1.1 Definitions and Elementary Properties

In [20] *J.P. La Salle* has developed a stability theory for difference equations. He considers difference equations which can be transformed into equations of the form

$$x(n+1) = f(x(n)), n \in \mathbb{N}_0 = \mathbb{N} \cup \{0\}, \qquad (1.1)$$

where

$$x(0) = x \qquad (1.2)$$

is a given initial state in a non-empty subset $\overline{X} \subseteq \mathbb{R}^k$ and $f : \overline{X} \to \overline{X}$ is a given continuous mapping. By (1.1) and (1.2) a time-discrete dynamical system $(\overline{X}, f)$ is defined, if we equip $\mathbb{R}^k$ with a norm (e.g. the *Euclidean norm*) and define a flow $\pi : \overline{X} \times \mathbb{N}_0 \to \overline{X}$ by

$$\pi(x,n) = f^n(x) = \underbrace{f \circ f \circ \ldots \circ f}_{n-times}(x) \qquad (1.3)$$

for all $x \in \overline{X}$ and $n \in \mathbb{N}$, and

$$\pi(x,0) = x, \text{ for all } x \in \overline{X}. \qquad (1.4)$$

This system is called an autonomous system, since it has the semi-group property

$$\pi(\pi(x,n),m) = \pi(x,n+m)$$

$$\text{for all } n, m \in \mathbb{N}_0 \text{ and } x \in \overline{X}.$$

The stability theory for such systems developed by J.P. La Salle can also be found in [13] in a slightly modified form. In this book we generalize the above situation as follows:

Let $\overline{X}$ be a metric space with metric $d : \overline{X} \times \overline{X} \to \mathbb{R}_+$ and let $f : \overline{X} \to \overline{X}$ be a continuous mapping. Then the autonomous time-discrete dynamical system $(\overline{X}, f)$ is given by the flow $\pi : \overline{X} \times \mathbb{N}_0 \to \overline{X}$ defined by (1.3) and (1.4). For every $x \in \overline{X}$ we define an orbit starting with x by

$$\gamma_f(x) = \bigcup_{n \in \mathbb{N}_0} \{f^n(x)\} . \tag{1.5}$$

Further we define as limit set of $\gamma_f(x)$ the set

$$L_f(x) = \bigcap_{n \in \mathbb{N}_0} \overline{\bigcup_{m \geq n} \{f^m(x)\}} . \tag{1.6}$$

This limit set can be given an equivalent definition which is the content of

Proposition 1.1. *For every $x \in \overline{X}$ the limit set $L_f(x)$ being defined by (1.6) consists of all accumulation points of the sequence $(f^n(x))_{n \in \mathbb{N}_0}$.*

Proof.

1) Let $y \in \overline{X}$ be an accumulation point of $(f^n(x))_{n \in \mathbb{N}_0}$. Then there exists a subsequence $(f^{n_i}(x))_{i \in \mathbb{N}_0}$ with $f^{n_i}(x) \to y$. This implies that, for every $n \in \mathbb{N}_0$,

$$y \in \overline{\bigcup_{m \geq n} \{f^m(x)\}}$$

which in turn implies that $y \in L_f(x)$.

2) Let $y \in L_f(x)$. Then

$$y \in \overline{\bigcup_{m \geq n} \{f^m(x)\}} \text{ for every } n \in \mathbb{N}_0.$$

Therefore, for every $n \in \mathbb{N}_0$, there is a sequence $(f^{k_i+n}(x))_{i \in \mathbb{N}_0}$ with $(f^{k_i+n}(x)) \to y$ as $i \to \infty$. Hence, for every $n \in \mathbb{N}_0$, there exists an $i_n \in \mathbb{N}_0$ such that

$$d(f^{k_i+n}(x), y) \leq \frac{1}{n} \text{ for all } i \geq i_n$$

and we can assume that $i_{n+1} > i_n$. This implies that $(f^{k_i+n}(x))_{n \in \mathbb{N}}$ is a subsequence of $(f^n(x))_{n \in \mathbb{N}_0}$ with $f^{k_i+n}(x) \to y$ as $n \to \infty$. This means that y is an accumulation point of $(f^n(x))_{n \in \mathbb{N}_0}$ which completes the proof.

$\square$

Definition 1.1. *A non-empty subset $H \subseteq \overline{X}$ is called positively (negatively) invariant (with respect to a mapping $f : \overline{X} \to \overline{X}$), if*

$$f(H) \subseteq H(\ H \subseteq f(H) \ ) \ ,$$

and invariant, if

$$f(H) = H \ .$$

Exercise 1.1. Show that for a continuous mapping $f : \overline{X} \to \overline{X}$ the following holds true:

(a) The closure of a positively invariant subset of $\overline{X}$ is also positively invariant (with respect to f).
(b) The closure of a relatively compact invariant subset of $\overline{X}$ is also invariant (with respect to f).

According to *Proposition 1.1* the limit set $L_f(x)$ for some $x \in \overline{X}$ given by (1.6) can be empty, if the sequence $(f^n(x))_{n \in \mathbb{N}_0}$ does not have accumulation points. If this is not the case, then we have the

Proposition 1.2. *If, for some $x \in \overline{X}$, the limit set $L_f(x)$ given by (1.6) is non-empty, then it is closed and positively invariant.*

Proof. The closedness of $L_f(x)$ is an immediate consequence of the Definition (1.6). Let $y \in L_f(x)$ be given. Then, by *Proposition 1.1*, there exists a subsequence $(f^{n_i}(x))_{i \in \mathbb{N}_0}$ of the sequence $(f^n(x))_{n \in \mathbb{N}_0}$ with $f^{n_i}(x) \to y$ as $i \to \infty$. This implies, due to the continuity of f, that $f^{n_i+1}(x) \to f(y)$, hence $f(y) \in L_f(x)$. This shows $f(L_f(x)) \subseteq L_f(x)$, i.e. that $L_f(x)$ is positively invariant.

□

If $\overline{X}$ is compact, then, for every $x \in \overline{X}$, the limit set $L_f(x)$ given by (1.6) is non-empty and we can even prove

Proposition 1.3. *If $\overline{X}$ is compact, then, for every $x \in \overline{X}$, the limit set $L_f(x)$ given by (1.6) is compact, invariant and the smallest closed subset $S \subseteq \overline{X}$ with*

$$\lim_{n \to \infty} \varrho(f^n(x), S) = 0 \text{ where } \varrho(y, S) = min\{d(y, z) \mid z \in S\}. \tag{1.7}$$

Proof. As a non-empty closed subset of the compact metric space $\overline{X}$ the limit set $L_f(x)$ is also compact.

In order to show the invariance of $L_f(x)$ it suffices to show that $L_f(x) \subseteq f(L_f(x))$. Let $y \in L_f(x)$ be given. Then there exists a subsequence $(f^{n_i}(x))_{i \in \mathbb{N}_0}$ of the sequence $(f^n(x))_{n \in \mathbb{N}_0}$ with $f^{n_i}(x) \to y$ and with no loss of generality we can assume that there is some $z \in L_f(x)$ with $f^{n_i-1}(x) \to z$. By the continuity of f this implies $f^{n_i}(x) \to f(z)$, hence $y = f(z) \in L_f(x)$ and therefore $L_f(x) \subseteq f(L_f(x))$. In order to show (1.7) we at first show that

$$\lim_{n \to \infty} \varrho(f^n(x), L_f(x)) = 0$$

For that purpose we assume that

$$\varrho(f^n(x), L_f(x)) \nrightarrow 0 \text{ as } n \to \infty.$$

Then there is a subsequence $(f^{n_i}(x))_{i \in \mathbb{N}_0}$ of $(f^n(x))_{n \in \mathbb{N}_0}$ with

$$\varrho(f^{n_i}(x), L_f(x)) \nrightarrow 0 \text{ as } i \to \infty$$

and some $y \in L_f(x)$ with $\lim_{i \to \infty} d(f^{n_i}(x), y) = 0$ which implies

$$\lim_{i \to \infty} \varrho(f^{n_i}(x), L_f(x)) = 0$$

and leads to a contradiction.

Now let $S \subseteq \overline{X}$ be any closed subset with

$$\lim_{n \to \infty} \varrho(f^n(x), S) = 0.$$

Then we choose any $y \in L_f(x)$ and conclude by *Proposition 1.1* the existence of a subsequence $(f^{n_i}(x))_{i \in \mathbb{N}_0}$ of $(f^n(x))_{n \in \mathbb{N}_0}$ with $(f_i^n(x)) \to y$. Further it follows that

$$\lim_{n \to \infty} \varrho(f^{n_i}(x), S) = 0,$$

which implies $y \in S$, since S is closed. This completes the proof of *Proposition 1.3.*

□

Definition 1.2. *A closed invariant subset of $\overline{X}$ is called invariantly connected, if it is not representable as a disjoint union of two non-empty, invariant and closed subsets of $\overline{X}$.*

Definition 1.3. *A sequence $(f^n(x))_{n \in \mathbb{N}_0}$, $x \in \overline{X}$, is called periodic or cyclic, if there exists a number $k \in \mathbb{N}$ with $f^k(x) = x$. The smallest number with this property is called the period of the sequence. If $k = 1$, then $x \in \overline{X}$ is called a fixed point of $f : \overline{X} \to \overline{X}$.*

Exercise 1.2. Show that a finite subset $H \subseteq \overline{X}$ is invariantly connected, if and only if for every $x \in H$ the sequence $(f^n(x))_{n \in \mathbb{N}_0}$ is periodic and its period is equal to the cardinality of H.

Proposition 1.3 can be supplemented by

Proposition 1.4. *If $\overline{X}$ is compact, then, for every $x \in \overline{X}$, the limit set $L_f(x)$ given by (1.6) is invariantly connected.*

Proof. Let us assume that, for some $x \in \overline{X}$, there exist two non-empty, invariant and closed subsets A_1 and A_2 of $L_f(x)$ which are disjoint and satisfy $A_1 \cup A_2 = L_f(x)$.
Since $L_f(x)$ is compact, A_1 and A_2 are also compact and there exist disjoint open subsets U_1 and U_2 of $\overline{X}$ with $A_1 \subseteq U_1$ and $A_2 \subseteq U_2$. Since f is uniformly continuous on A_1, there is an open subset V_1 of $\overline{X}$ with $A_1 \subseteq V_1$ and $f(V_1) \subseteq U_1$.
Since $L_f(x)$ is the smallest closed set $S \subseteq \overline{X}$ with (1.7), the sequence $(f^n(x))_{n \in \mathbb{N}_0}$ must intersect V_1 as well as U_2 infinitely many times. This implies the existence of a subsequence $(f^{n_i}(x))_{i \in \mathbb{N}_0}$ which is neither contained in V_1 nor U_2 and which can be assumed to be convergent. This, however, is impossible and leads to a contradiction to the assumption that $L_f(x)$ is not invariantly connected.

$\square$

Exercise 1.3. Given a compact and positively invariant subset $K \subseteq \overline{X}$. Show that $\bigcap_{n \in \mathbb{N}_0} f^n(K)$ with $f^n(K) = \{f^n(x) \mid x \in K\}$ for all $n \in \mathbb{N}_0$ is non-empty, compact and the largest invariant subset of K.

1.1.2 Localization of Limit Sets with the Aid of Lyapunov Functions

Let $\overline{X}$ be a metric space and let $f : \overline{X} \to \overline{X}$ be a continuous mapping. Further let $G \subseteq \overline{X}$ be a non-empty subset.

Definition 1.4. *A function* $V : \overline{X} \to \mathbb{R}$ *is called a* Lyapunov function *with respect to* f *on* G, *if*

(1) V *is continuous on* $\overline{X}$;
(2) $V(f(x)) - V(x) \leq 0$ *for all* $x \in G$ *with* $f(x) \in G$.

If $V : \overline{X} \to \mathbb{R}$ *is a Lyapunov function with respect to* f *on* G, *then we define the set*

$$E = \{x \in \overline{X} \mid V(f(x)) = V(x), x \in \overline{G}\}$$

where $\overline{G}$ *is the closure of* G. *Further we put, for every* $c \in \mathbb{R}$,

$$V^{-1}(c) = \{x \in \overline{X} \mid V(x) = c\}.$$

Then we can prove

Proposition 1.5. *Let* $G \subseteq \overline{X}$ *be non-empty and relatively compact. Further let* V *be a Lyapunov function with respect to* f *on* G *and finally let* $x_0 \in G$ *be such that*

$$f^n(x_0) \in G \text{ for all } n \in \mathbb{N}.$$

Then there exists some $c \in \mathbb{R}$ *such that*

$$L_f(x_0) \subseteq M \cap V^{-1}(c)$$

where M *is the largest invariant subset of* E.

Proof. If we define $x_n = f^n(x_0)$, $n \in \mathbb{N}_0$, then the sequence $(V(x_n)_{n \in \mathbb{N}_0})$ is contained in $V(\overline{G})$ and therefore bounded.
Further it follows that

$$V(x_{n+1}) \leq V(x_n) \text{ for all } n \in \mathbb{N}_0.$$

Therefore there exists some $c \in \mathbb{R}$ with $c = \lim_{n \to \infty} V(x_n)$.
Now let $p \in L_f(x_0)$. Then there exists a subsequence $(x_{n_i})_{i \in \mathbb{N}_0}$ of $(x_n)_{n \in \mathbb{N}_0}$ with $(x_{n_i}) \to p$ which implies $c = \lim_{i \to \infty} V(x_{n_i}) = V_p$ due to the continuity of V. Hence $p \in V^{-1}(c)$ and thus $L_f(x_0) \subseteq V^{-1}(c)$. Since, by *Proposition 1.3*, $L_f(x_0)$ is invariant, it follows that $V(f(p)) = c$ for all $p \in L_f(x_0)$ and hence $V(f(p)) = V(p)$ for all $p \in L_f(x_0)$ which implies $L_f(x_0) \subseteq E$ and in turn $L_f(x_0) \subseteq M$. This completes the proof.

$\square$

Let us demonstrate this result by an example: We take $\underline{X} = \mathbb{R}^2$ equipped with the *Euclidean norm* and define $f : \underline{X} \to \underline{X}$ by

$$f(x,y) = (f_1(x,y) \, , \; f_2(x,y))^T \, , \; (x,y) \in \underline{X} \, ,$$

with

$$f_1(x,y) = \frac{y}{1+x^2} \, , \; f_2(x,y) = \frac{x}{1+y^2} \, . \tag{1.8}$$

Further we choose

$$V(x,y) = x^2 + y^2 \, , \; (x,y) \in \underline{X} \, . \tag{1.9}$$

Then it follows that

$$V\left(f(x,y)\right) - V(x,y) = \left(\frac{1}{(1+y^2)^2)} - 1\right) x^2 + \left(\frac{1}{(1+x^2)^2} - 1\right) y^2 \le 0 \, ,$$

$$\text{for all } (x,y) \in \underline{X}$$

This shows that V is a *Lyapunov function* with respect to f on $\underline{X} = \mathbb{R}^2$. Further we see that

$$E = \{(x,0) \mid x \in \mathbb{R}\} \cup \{(0,y) \mid y \in \mathbb{R})\}.$$

From $f(x,0) = (0,x)$ for all $x \in \mathbb{R}$ and $f(0,y) = (y,0)$ for all $y \in \mathbb{R}$, it follows that E is invariant and hence $M = E$.
Further we conclude

$$f^2(x,0) = f(0,x) = (0,x) \text{ for all } x \in \mathbb{R}$$

and

$$f^2(0,y) = f(y,0) = (0,y) \text{ for all } x \in \mathbb{R}.$$

Now let $G = \{(x,y) \in \mathbb{R}^2 \mid x^2 + y^2 < r\}$ for any $r > 0$.
Then it follows that $f(G) \subseteq G$ and for every $(x_0, y_0) \in G$ there exists some $c \in \mathbb{R}$ such

$$L_f(x_0,y_0) \subseteq E \cap \{(x,y) \mid x^2 + y^2 = c^2\} = \{(c,0),(0,c)\}.$$

Since $L_f(x_0,y_0)$ is invariantly connected by *Proposition 1.4* it follows from *Exercise 1.2* that

$$L_f(x_0,y_0) = \{(c,0),(0,c)\}.$$

1.1.3 Stability Based on Lyapunov's Method

Let $f : \overline{X} \to \overline{X}$ be a continuous mapping where $\overline{X}$ is a metric space.

Definition 1.5. *A relatively compact set $H \subseteq \overline{X}$ is called stable with respect to f, if for every relatively compact open set $U \subseteq \overline{X}$ with $U \supseteq \overline{H} = $ **closure of** H there exists an open set $W \subseteq \overline{X}$ with $\overline{H} \subseteq W \subseteq U$ such that*

$$f^n(W) \subseteq U \text{ for all } n \in \mathbb{N}_0$$

where

$$f^n(W) = \{f^n(x) \mid x \in W\}.$$

Theorem 1.1. *Let $H \subseteq \overline{X}$ be relatively compact and such that for every relatively compact open set $U \subseteq \overline{X}$ with $U \supseteq \overline{H}$ there exists an open set subset B_U of U with $B_U \supseteq \overline{H}$ and $f(B_U) \subseteq B_U$.*
Further let $G \subseteq \overline{X}$ be an open set with $G \supseteq \overline{H}$ such that there exists a Lyapunov function V with respect to f on G which is positive definite with respect to $\overline{H}$, i.e.,

$$V(x) \geq 0 \text{ for all } x \in G \text{ and } (V(x) = 0 \Leftrightarrow x \in \overline{H}).$$

Then H is stable with respect to f.

Proof. Let $U \subseteq \overline{X}$ be an arbitrary relatively compact open set with $U \supseteq \overline{H}$. Then $U^* = U \cap G$ is also a relatively compact open set with $U^* \supseteq \overline{H}$ and there exists an open set $B_{U^*} \subseteq U^*$ with $B_{U^*} \supseteq \overline{H}$ and $f(B_{U^*}) \subseteq U^*$.
Let us put

$$m = min\{V(x) \mid x \in \overline{U^*} \backslash B_{U^*}\}.$$

Since $\overline{H} \cap (\overline{U^*} \backslash B_{U^*})$ is empty, it follows that $m > 0$. If we define

$$W = \{x \in U^* \mid V(x) < m\},$$

then W is open and $\overline{H} \subseteq W \subseteq B_{U^*}$.
Now let $x \in W$ be chosen arbitrarily. Then $x \in B_{U^*}$ and therefore $f(x) \in U^*$. Further we have

$$V(f(x)) \leq V(x) < m,$$

hence $f(x) \in W \subseteq B_{U^*}$. This implies $f^2(x) = f(f(x)) \in U^*$ and

$$V(f^2(x)) \leq V(f(x)) < m, \text{ hence } f^2(x) \in W.$$

By induction it therefore follows that

$$f^n(x) \in W \subseteq U^* \subseteq U \text{ for all } n \in \mathbb{N}_0.$$

This shows that H is stable with respect to f.

$\square$

Exercise 1.4. (a) Give an explicit definition of the instability of a relatively compact set $H \subseteq \overline{X}$ with respect to f as logical negation of the notion of stability. (b) Show: If a relatively compact set $H \subseteq \overline{X}$ is stable with respect to f, then its closure $\overline{H}$ is positively invariant, i.e. $f(\overline{H}) \subseteq \overline{H}$.

Definition 1.6. *A set $H \subseteq \overline{X}$ is called an attractor with respect to f, if there exists an open set $U \subseteq \overline{X}$ with $U \supseteq \overline{H}$ such that*

$$\lim_{n \to \infty} \varrho(f^n(x), \overline{H}) = 0 \text{ (in short: } f^n(x) \to \overline{H}) \text{ for all } x \in U$$

where

$$\varrho(y, \overline{H}) = \inf\{d(y, z) \mid z \in \overline{H}\}.$$

If $H \subseteq \overline{X}$ is stable and an attractor with respect to f, then H is called asymptotically stable with respect to f.

Theorem 1.2. *Let $H \subseteq \overline{X}$ be such that there exists a relatively compact open set $U \supseteq \overline{H}$ with $f(U) \subseteq U$.*
Further let $V : \overline{X} \to \mathbb{R}$ be a Lyapunov function with respect to f on U which is positive definite with respect to $\overline{H}$, i.e.

$$V(x) \geq 0 \text{ for all } x \in U \text{ and } (V(x) = 0 \Leftrightarrow x \in \overline{H}).$$

Finally let

$$\lim_{n \to \infty} V(f^n(x)) = 0 \text{ for all } x \in U.$$

Then H is an attractor with respect to f.

Proof. Let $x \in U$ be chosen arbitrarily. Since the sequence $(f^n(x))_{n \in \mathbb{N}_0}$ is contained in U, for every subsequence $(f^{n_i}(x))_{i \in \mathbb{N}_0}$ of $(f^n(x))_{n \in \mathbb{N}_0}$ there exists a subsequence $(f^{n_{i_j}}(x))_{j \in \mathbb{N}_0}$ and some $q \in U$ with

$$\lim_{j \to \infty} f^{n_{i_j}}(x) = q.$$

This implies

$$\lim_{j \to \infty} V(f^{n_{i_j}}(x)) = V(q) = 0,$$

hence $q \in \overline{H}$, and therefore $f^{n_{i_j}}(x) \to \overline{H}$ as $j \to \infty$. From this it follows that $f^n(x) \to \overline{H}$ which shows that H is an attractor with respect to f.

$\square$

Corollary: *Under the assumptions of Theorem 1.1 and 1.2 it follows that $H \subseteq \overline{X}$ is asymptotically stable.*

As an important special case we prove the following

Theorem 1.3. *Let $x^* \in \overline{X}$ be a fixed point of f, i.e. $f(x^*) = x^*$. Further let there exist an open set $G \subseteq \overline{X}$ with $x^* \in G$ and a Lyapunov function V with respect to f on G which is positive definite with respect to x^*, i.e.*

$$V(x) \geq 0 \text{ for all } x \in G \text{ and } (V(x) = 0 \Leftrightarrow x = x^*).$$

Then $\{x^*\}$ is stable with respect to f.
If in addition

$$V(f(x)) < V(x) \text{ for all } x \in G \text{ with } x \neq x^* , \qquad (1.10)$$

then $\{x^*\}$ is asymptotically stable with respect to f.

Proof. Let $U \subseteq \overline{X}$ be a relatively compact open set with $x^* \in U$. Then there exists some $r > 0$ such that

$$B_r(x^*) = \{x \in \overline{X} \mid d(x, x^*) < r\} \subseteq U.$$

Since f is continuous in x^*, there exists some $s \in (0, r)$ such that $f(B_s(x^*)) \subseteq B_r(x^*)$. Hence, if we put $B_U = B_s(x^*)$, then $f(B_U) \subseteq U$ is open and $\{x^*\} \subseteq B_U$. By *Theorem 1.1* therefore $\{x^*\}$ is stable with respect to f.
Now let U be any relatively compact subset of G with $x^* \in U$.
Then we define

$$q = \sup_{x \in U} \frac{V(f(x))}{V(x)}$$

and conclude from (1.10) that $0 < q < 1$.
Further it follows that

$$V(f^n(x)) \leq q^n V(x) \text{ for all } x \in u \text{ and } n \in \mathbb{N}.$$

This implies

$$\lim_{n \to \infty} V(f^n(x)) = 0 \text{ for all } x \in U.$$

Hence by *Theorem 1.2* we conclude that $\{x^*\}$ is an attractor and therefore asymptotically stable with respect to f.

$\square$

Let us demonstrate this result by the same example as in *Section 1.1.2*, i.e. we choose $f : \mathbb{R}^2 \to \mathbb{R}^2$ as

$$f(x, y) = (f_1(x, y), f_2(x, y)), (x, y) \in \overline{X} = \mathbb{R}^2,$$

with $f_1 = f_1(x, y)$ and $f_2 = f_2(x, y)$ defined by (1.8).

If we choose $V : \overline{X} \to \mathbb{R}$ in the form of (1.9), then V is a *Lyapunov function* with respect to f on $G = \overline{X} = \mathbb{R}^2$.

If we choose $x^* = (0, 0)^T$, then V is positive definite with respect to x^* and x^* is a fixed point of f. Further (1.10) is satisfied.

By *Theorem 1.3* it follows that $x^* = (0, 0)^T$ is asymptotically stable with respect to f.

With the aid of *Proposition 1.5* we can prove the

Theorem 1.4. *Let $G \subseteq \overline{X}$ be open, relatively compact and positively invariant with respect to f, i.e. $f(G) \subseteq G$. Further let V be a Lyapunov function with respect to f on G. Then the largest invariant subset M of*

$$E = \{x \in \overline{G} \mid V(f(x)) = V(x)\}$$

is an attractor with respect to f.
If in addition V is constant on M, then M is asymptotically stable with respect to f.

Proof. Let us put $U = G$. If we choose $x_0 \in U$ arbitrarily, then the sequence $(f^n(x_0))_{n \in \mathbb{N}_0}$ is contained in G, since G is positively invariant. *Proposition 1.5* implies $L_f(x_0) \subseteq M$. We have to show that $\lim_{n \to \infty} \varrho(f^n(x_0), \overline{M} = 0)$. Now let $(f^{n_i}(x_0))_{i \in \mathbb{N}_0}$ be an arbitrary subsequence of $(f^n(x_0))_{n \in \mathbb{N}_0}$. Then there is a subsequence $(f^{n_{ij}}(x_0))_{j \in \mathbb{N}_0}$ and an element $y \in L_f(x_0) \subseteq M$ with $f^{n_{ij}}(x_0) \to y$ which implies $\lim_{j \to \infty} \varrho(f^{n_j}(x_0), \overline{M}) = 0$ and in turn $\varrho(f^n(x_0), \overline{M}) \to y$.

In order to show that M is asymptotically stable with respect to f we have to show that M is stable with respect to f. For that purpose we apply *Theorem 1.1* and verify its assumptions. At first let $U \subseteq \overline{X}$ be relatively compact, open and such that $U \supseteq \overline{H}$. If we define

$$B_U = \{x \in U \cap G \mid f(x) \in U\},$$

then B_U is open, $B_U \supseteq \overline{M}, B_U \subseteq U$, and $f(B_U) \subseteq U$.

Let

$$V(x) = c \ \text{ for all } \ x \in M \ .$$

(1.11)

Let us assume that there is some $x \in G$ with $V(x) < c$.
Then it follows that

$$V(f^n(x)) \leq V(x) < c \ \text{ for all } \ n \in \mathbb{N}.$$

Since $L_f(x) \subseteq M$, there exists some $y \in L_f(x) \subseteq M$ with $V(y) < c$ which
contradicts (1.11). Hence

$$V(x) \geq c \text{ for all } x \in G$$

and further

$$V(x) = c \Leftrightarrow x \in \overline{M}.$$

Therefore, if we define

$$\tilde{V}(x) = V(x) - c, \quad x \in \overline{X},$$

then we obtain a *Lyapunov function* with respect to f on G which is positive
definite with respect to $\overline{M}$. The assertion now follows by *Theorem 1.1*.

□

Let us again demonstrate this result by the above example (1.8) with *Lyapunov function* (1.9). Let again

$$G = \{(x,y) \in \mathbb{R}^2 \mid x^2 + y^2 < r^2\} \text{ for some } r > 0.$$

Then G is open, relatively compact and positively invariant.
From *Theorem 1.4* it therefore follows that

$$M = (\{(x,0) \mid x \in \mathbb{R}\} \cup \{(0,x) \mid x \in \mathbb{R}\}) \cap G$$

is an attractor with respect to f. It even follows that for every $x_0 \in G$ it is
true that

$$\lim_{n \to \infty} \varrho(f^n(x_0), \overline{M}) = 0.$$

1.1.4 Stability of Fixed Points via Linearisation

Let us assume that $\overline{X}$ is a non-empty open subset of a normed linear space $(E, \|\cdot\|)$ and $f : \overline{X} \to \overline{X}$ is a continuous mapping which is continuously *Fréchet* differentiable at every $x \in \overline{X}$. We denote the *Fréchet* derivative of f at $x \in \overline{X}$ by f'_x which is a continuous linear mapping $f'_x : E \to E$ whose norm is given by

$$\|f'_x\| = \sup\{ \|f'_x(h)\| \mid h \in E \text{ with } \|h\| = 1 \}$$

for $x \in \overline{X}$. Then we can prove

Theorem 1.5. *Let $x^* \in \overline{X}$ be a fixed point of f, i.e. $f(x^*) = x^*$. Then the following two statements are true:*

a) *Let $\|f'_{x^*}\| < 1$. Then there is some $\varepsilon > 0$ and some $c \in [0,1)$ such that*

$$x_0 \in \overline{X} \text{ and } \|x_0 - x^*\| < \varepsilon \Rightarrow \|f^n(x_0) - x^*\| \leq c^n \|x_0 - x^*\| \text{ for all } n \in \mathbb{N}$$

which implies that x^* is asymptotically stable with respect to f (Exercise).

b) *Let f'_{x^*} be continuously invertible and $\|f'_{x^*}{}^{-1}\|^{-1} > 1$. Then there is some $\delta > 0$ and some $d > 1$ such that*

$$f^k(x_0) \in \overline{X} \text{ and } \|f^k(x_0) - x^*\| < \delta \text{ for } 0 \leq k \leq n - 1$$

implies that

$$\|f^n(x_0) - x^*\| \geq d^n \|x_0 - x^*\|$$

for all $n \in \mathbb{N}$. This implies that x^* is unstable with respect to f (Exercise, see Exercise 1.4 a)).

Proof. a) From the continuity of $x \to \|f'_x\|$, $x \in \overline{X}$, it follows that there exists some $\varepsilon > 0$ and some $c \in [0,1)$ such that

$$\|f'_x\| \leq c \text{ for all } x \in B(x^*, \varepsilon) = \{y \in E \mid \|y - x^*\| < \varepsilon\}.$$

The mean value theorem implies

$$\|f(x) - f(y)\| \leq \|f'_z\| \, \|x - y\|$$

for all $x, y \in B(x^*, \varepsilon)$ and some $z = \alpha x + (1 - \alpha)y$ with $\alpha \in (0, 1)$. This implies

$$\|f(x) - f(y)\| \leq c\|x - y\| \text{ for all } x, y \in B(x^*, \varepsilon)$$

and in particular

$$\|f(x) - x^*\| \leq c\|x - x^*\| < \varepsilon \text{ for all } x \in B(x^*, \varepsilon).$$

Therefore it follows that $f(B(x^*, \varepsilon)) \subseteq B(x^*, \varepsilon)$ and hence

$$\|f^n(x_0) - x^*\| = \|f \circ f^{n-1}(x_0) - f(x^*)\| \leq c\|f^{n-1}(x_0) - x^*\|$$

$$\leq \ldots \leq c^n \|x_0 - x^*\| \text{ for all } x_0 \in B(x^*, \varepsilon) \text{ and } n \in \mathbb{N}$$

b) For every $x, y \in \overline{X}$ we have

$$\|x - y\| = \|f_{x^*}'^{-1}(f_{x^*}'(x)) - f_{x^*}'^{-1}(f_{x^*}'(y))\| \leq \|f_{x^*}'^{-1}\| \, \|f_{x^*}'(x - y)\|$$

which implies

$$\|f_{x^*}'(x - y)\| \geq d' \|x - y\| \quad \text{with}$$

$d' = \|f_{x^*}'^{-1}\|^{-1} > 1$. From the *Fréchet differentiability* it follows that

$$f(x) - f(x^*) = f_{x^*}'(x - x^*) + \varepsilon(\|x - x^*\|) \text{ for } x \in \overline{X}$$

where

$$\lim_{x \to x^*} \frac{\|\varepsilon(\|x - x^*\|)\|}{\|x - x^*\|} = 0.$$

If one chooses $\eta > 0$ with $d = d' - \eta > 1$ and $\delta > 0$ such that

$$\frac{\|\varepsilon(\|x - x^*\|)\|}{\|x - x^*\|} \leq \eta \quad \text{for all} \quad B(x^*, \delta) \setminus \{x^*\} \, ,$$

then it follows that

$$\|f(x) - f(x^*)\| \geq \|f_{x^*}'(x - y)\| - \|\varepsilon(\|x - x^*\|)\|$$

$$\geq d' \|x - x^*\| - \eta \|x - x^*\|$$

$$= d \|x - x^*\|$$

for all $x \in B(x^*, \delta)$. Now let $x_0 \in B(x^*, \delta)$ be such that

$$\|f^k(x_0) - x^*\| < \delta \text{ for } 0 \leq k \leq n - 1 \, , \, n \in \mathbb{N}.$$

Then it follows that

$$\|f^n(x_0) - x^*\| = \|f \circ f^{n-1}(x_0) - f(x^*)\| \geq d \|f^{n-1}(x_0) - x^*\|$$

$$\geq \ldots \geq d^n \|x_0 - x^*\|.$$

This completes the proof of *Theorem 1.5*.

$\square$

Let us consider the following important special case:

We assume $E = \mathbb{R}^k$ equipped with any norm $\| \cdot \|$. Let again $\overline{X} \subseteq E$ be non-empty and open and let again $f : \overline{X} \to \overline{X}$ be a continuously *Fréchet differentiable* (and hence continuous) mapping on $\overline{X}$. This is equivalent to the statement that at every $x \in \overline{X}$ there exists the *Jacobian matrix*

$$J_f(x) = \begin{pmatrix} f_{1x_1}(x) & \cdots & f_{1x_k}(x) \\ \vdots & \ddots & \vdots \\ f_{kx_1}(x) & \cdots & f_{kx_k}(x) \end{pmatrix} , \; x = (x_1, \ldots, x_k),$$

and depends continuously on x.
The *Fréchet derivative* is then given by

$$f'_x(h) = J_f(x)h \; , \; h \in \mathbb{R}^k,$$

and we obtain

$$\|f'_x\| = \|J_f(x)\| = \sup\{ \; \|J_f(x)(h)\| \; | \; \|h\| = 1 \; \}.$$

Theorem 1.5 now gives rise to the following

Corollary: *Let $x^* \in \overline{X}$ be a fixed point of f. Then the following two statements are true:*

a) *Let the spectral radius $\varrho(J_f(x^*)) < 1$. Then x^* is asymptotically stable with respect to f.*
b) *Let $J_f(x^*)$ be invertible and let all the eigenvalues of $J_f(x^*)$ be larger than 1 in absolute value. Then x^* is unstable with respect to f.*

Proof. By *Theorem 3* in *Chapter 1* of the book *"Analysis of Numerical Methods"* by E. Isaacson and H.B. Keller *(John Wiley and Sons, New York, London, Sydney 1966)* there exists, for every $\delta > 0$, a vector norm in $\mathbb{R}^k$ such that

$$\|J_f(x^*)\| \leq \varrho(J_f(x^*)) + \delta.$$

a) Let us choose

$$\delta = \frac{1}{2}(1 - \varrho(J_f(x^*)))(> 0).$$

Then

$$\|f'_{x^*}\| = \|J_f(x^*)\| \leq \frac{1}{2}(1 + \delta(J_f(x^*))) < 1$$

and the assertion follows from *Theorem 1.5 a)*.

b) From the assumption it follows that $\varrho(J_f(x^*)^{-1}) < 1$. Again by the above quoted theorem it follows that, for a suitable matrix norm,

$$\|J_f(x^*)^{-1}\| < 1 \Rightarrow 1 < \|J_f(x^*)^{-1}\|^{-1} = \|f_{x^*}'^{-1}\|^{-1}$$

so that the assertion follows from *Theorem 1.5 b)*.

□

Assumption *a)* in the *Corollary* is not necessary for the asymptotic stability of x^* with respect to f as can be seen by the Example (1.8). Here we have

$$f_{1x}(x,y) = -\frac{2xy}{(1+x^2)^2} \ , \quad f_{1y}(x,y) = \frac{1}{1+x^2} \ ,$$

$$f_{2x}(x,y) = \frac{1}{1+y^2} \ , \quad f_{2y}(x,y) = -\frac{2xy}{(1+y^2)^2} \ ,$$

hence

$$J_f(0,0) = \begin{pmatrix} 0 & 1 \\ 1 & 0 \end{pmatrix} \Rightarrow \varrho(J_f(0,0)) = 1.$$

However, we have seen in *Section 1.1.3* that $(0,0)$ is asymptotically stable with respect to f.

1.1.5 Linear Systems

As at the beginning of *Section 1.1.4* we consider a normed linear space $(E, \|\cdot\|)$ and a mapping $f : E \to E$ which is given by

$$f(x) = A(x) + b \ , \ x \in E,$$

where $A : E \to E$ is a continuous linear mapping and $b \in E$ a fixed element. Then f is *Fréchet differentiable* at every $x \in E$ and its *Fréchet derivative* is given by

$$f_x' = A \text{ for all } x \in E.$$

Theorem 1.5 leads immediately to

Theorem 1.6. *Let* $x^* \in E$ *be a fixed point of* f, *i.e.*

$$x^* = A(x^*) + b. \tag{1.12}$$

Then the following two statements are true:

a) *If* $\|A\| < 1$, *then* x^* *is asymptotically stable with respect to* f.
b) *Let* A *be continuously invertible and* $\|A^{-1}\|^{-1} > 1$. *Then* x^* *is unstable with respect to* f.

Exercise 1.5. Show that under the assumption $\|A\| < 1$ there is at most one $x^* \in E$ with *(1.12)*.
Further show that, if $\|A\| < 1$ and there exists $x^* \in E$ with *(1.12)* (which is then unique), it follows that, for every $x_0 \in E$, the sequence $(x_n)_{n \in \mathbb{N}_0}$ given by

$$x_{n+1} = A(x_n) + b, \, , \, n \in \mathbb{N}_0,$$

converges to x^*.

The Corollary of *Theorem 1.5* leads to the following
Corollary: *Let* $E = \mathbb{R}^k$ *equipped with any norm and let*

$$f(x) = Ax + b \, , \, x \in \mathbb{R}^k,$$

where A *is a real* $k \times k - matrix$ *and* $b \in \mathbb{R}^k$ *a fixed element. Let* $x^* \in \mathbb{R}^k$ *be a fixed point of* f, *i.e.*

$$x^* = Ax^* + b.$$

Then the following two statements are true:

a) *If* $\varrho(A) < 1$, *then* x^* *is asymptotically stable with respect to* f.
b) *If* A *is invertible and all eigenvalues of* A *are larger than one in absolute value, then* x^* *is unstable with respect to* f.

In the following we introduce a concept of stability and asymptotic stability for linear systems that differs from the one given in *Section 1.1.3* and that we adopt from [18]. For that purpose we consider the sequences $(x_n)_{n \in \mathbb{N}_0}$ in E with

$$x_{n+1} = Ax_n + b \, , \, n \in \mathbb{N}_0, x_0 \in E. \tag{1.13}$$

Definition 1.7. *A sequence* $(x_n)_{n \in \mathbb{N}_0}$ *in* E *with* (1.13) *is called*

1. stable, *if for every* $\varepsilon > 0$ *and every* $N \in \mathbb{N}$, *there exists some* $\delta = \delta(\varepsilon, N)$ *such that for every sequence* $(\bar{x}_n)_{n \in \mathbb{N}_0}$ *with* (1.13) *for* $\bar{x}_0 \in E$ *and* $\|\bar{x}_N - x_N\| < \delta$ *it follows that*

$$\|\bar{x}_n - x_n\| < \varepsilon \text{ for all } n \geq N;$$

2. attractive, *if for every* $N \in \mathbb{N}$, *there exists some* $\delta = \delta(N)$ *such that for every sequence* $(\bar{x}_n)_{n \in \mathbb{N}_0}$ *with* (1.13) *for* $(\bar{x}_0) \in E$ *and* $\|\bar{x}_N - x_N\| < \delta$ *it follows that* $\lim_{n \to \infty} \|\bar{x}_n - x_n\| = 0$;

3. asymptotically stable, *if* $(x_n)_{n \in \mathbb{N}_0}$ *is stable and attractive*.

In order to guarantee the stability of all sequences $(x_n)_{n \in \mathbb{N}_0}$ with (1.13) it suffices to guarantee the stability of the zero sequence $(x_n = \Theta_E)_{n \in \mathbb{N}_0}$ which satisfies (1.13) for $b = \Theta_E =$ zero element of E. This is a consequence of

Lemma 1.1. *The following statements are equivalent:*

(1) All sequences $(x_n)_{n \in \mathbb{N}_0}$ with (1.13) are stable.
(2) One sequence $(x_n)_{n \in \mathbb{N}_0}$ with (1.13) is stable.
(3) The zero sequence $(x_n = \Theta_E)_{n \in \mathbb{N}_0}$ which satisfies (1.13) for $b = \Theta_E$ is stable.

Proof. $(1) \Rightarrow (2)$ is trivially true. Now let (2) be true. Then there is a sequence $(y_n)_{n \in \mathbb{N}_0}$ with (1.13) which is stable. Now let $(\overline{x}_n)_{n \in \mathbb{N}_0}$ be any sequence with (1.13) for $b = \Theta_E$. Then $(\overline{x}_n + y_n)_{n \in \mathbb{N}_0}$ satisfies (1.13) and for every $\varepsilon > 0$ there exists some $\delta = \delta(\varepsilon, N)$ such that

$$\|\overline{x}_N - \Theta_E\| = \|\overline{x}_N + y_N - y_N\| < \delta \text{ implies}$$

$$\|\overline{x}_n - \Theta_E\| = \|\overline{x}_n + y_n - y_n\| < \varepsilon \text{ for all } n \geq N,$$

since $(y_n)_{n \in \mathbb{N}}$ is stable. From this it follows that $(x_n = \Theta_E)_{n \in \mathbb{N}_0}$ is stable which shows $(2) \Rightarrow (3)$.
Now let $(y_n)_{n \in \mathbb{N}_0}$ and $(\overline{y}_n)_{n \in \mathbb{N}_0}$ be arbitrary sequences with (1.13). Then we choose any sequence $(z_n)_{n \in \mathbb{N}_0}$ in E with (1.13) and define

$$x_n = y_n - z_n \text{ and } \overline{x}_n = \overline{y}_n - z_n \text{ for } n \in \mathbb{N}_0.$$

Since according to (3) the zero sequence $(\tilde{x}_n = \Theta_E)_{n \in \mathbb{N}_0}$ (which satisfies (1.13) for $b = \Theta_E$) is stable, it follows that for every $\varepsilon > 0$ there exists some $\delta = \delta(\varepsilon, N)$ such that

$$\|y_N - \overline{y}_N\| = \|x_N - \overline{x}_N - \Theta_E\| < \delta \text{ implies that}$$

$$\|y_n - \overline{y}_n\| = \|x_n - \overline{x}_n - \Theta_E\| < \varepsilon \text{ for all } n \geq N$$

which shows that (1) is true.
Hence $(1) \Rightarrow (2) \Rightarrow (3) \Rightarrow (1)$ which completes the proof.

$\square$

Remark: *Lemma 1.1* also holds true, if we replace stable by attractive. Hence it is also true with asymptotically stable instead of stable.

Now let us again consider the special case $E = \mathbb{R}^k$ equipped with any norm $\|\cdot\|$. According to *Lemma 1.1* we consider sequences $(x_n)_{n \in \mathbb{N}_0}$ in $\mathbb{R}^k$ with

$$x_{n+1} = Ax_n , n \in \mathbb{N}_0 , x_0 \in \mathbb{R}^k. \tag{1.14}$$

In order to show stability of the zero sequence $(x_n = \Theta_n)_{n \in \mathbb{N}_0}$ we assume that A has *eigenvalues* $\lambda_1, \ldots, \lambda_k \in \mathbb{C}$ such that there exist *eigenvectors* $v_1, \ldots, v_k \in \mathbb{C}^k$ which are linearly independent. Therefore every $x_0 \in \mathbb{R}^k$ has a unique representation

$$x_0 = \sum_{i=1}^{k} c_i v_i \ , \ c_1, \ldots, c_k \in \mathbb{C}.$$

This implies that every sequence $(x_n)_{n \in \mathbb{N}_0}$ can be represented in the form

$$x_n = \sum_{i=1}^{k} c_i \lambda_i^n v_i \ , \ n \in \mathbb{N}_0.$$

Now let us assume that

$$|\lambda_i| \leq 1 \text{ for } i = 1, \ldots, k. \tag{1.15}$$

If we define, for every

$$z = \sum_{i=1}^{k} c_i(z) v_i \in \mathbb{C}^k \ ,$$

a norm by

$$\|z\| = \sum_{i=1}^{k} |c_i(z)| \ ,$$

it follows that

$$\|x_n\| = \sum_{i=1}^{k} |\lambda_i|^n \ |c_i(x_0)|$$

$$\leq \sum_{i=1}^{k} |c_i(x_0)| = \|x_0\| \ .$$

This leads to

Theorem 1.7. *Let the eigenvalues $\lambda_1, \ldots, \lambda_k \in \mathbb{C}$ of A satisfy (1.15) and be such that the corresponding eigenvectors are linearly independent. Then the zero sequence $(x_n = \Theta_k)_{n \in \mathbb{N}_0}$ which satisfies (1.14) is stable and hence every sequence $(x_n)_{n \in \mathbb{N}_0}$ that satisfies (1.14) with $E = \mathbb{R}^k$ is stable.*

Proof. Let $\varepsilon > 0$ be chosen. Then we put $\delta = \varepsilon$. Now let $(\bar{x}_n)_{n \in \mathbb{N}_0}$ be any sequence with (1.14) for $x_n = \bar{x}_n$ and $\|\bar{x}_0\| < \delta$. Then it follows that $\|\bar{x}_n - \Theta_k\| < \varepsilon$ for all $n \in \mathbb{N}_0$ which completes the proof.

$\square$

Exercise 1.6. Prove that the zero sequence $(x_n = \Theta_k)_{n \in \mathbb{N}_0}$ is asymptotically stable, if all the *eigenvalues* of A are less than 1 in absolute value.

Remark: Under the assumptions of *Theorem 1.7* it is also true that the mapping $f : \mathbb{R}^k \to \mathbb{R}^k$ defined by

$$f(x) = Ax , \; x \in \mathbb{R}^k ,$$

has $\{\Theta_k\}$ as a stable fixed point in the sense of *Definition 1.5*.

Proof. Let $G = \mathbb{R}^k$ and define $V : \mathbb{R} \to \mathbb{R}$ by

$$V(x) = \|x\| , x \in \mathbb{R}^k ,$$

where

$$\|x\| = \sum_{i=1}^{k} |c_i(x)| , \; x \in \mathbb{R}^k .$$

Then it follows that

$$V(Ax) = \|Ax\| = \sum_{i=1}^{k} |\lambda_i| \, |c_i(x)| \le \sum_{i=1}^{k} |c_i(x)| = \|x\| = V(x) , \; x \in \mathbb{R}^k .$$

Hence $V(Ax) - V(x) \le 0$ for all $x \in G$ and V is a Lyapunov function on $G = \mathbb{R}^k$.

Further we have

$$V(x) \ge 0 \quad \text{for all } x \in G \text{ and } (V(x) = 0 \Leftrightarrow x = \Theta_k) .$$

Theorem 1.3 therefore implies that $\{\Theta_k\}$ is a stable fixed point of f.

$\square$

We can also use another *Lyapunov function* in order to show that $\hat{x} = \Theta_k$ is a stable fixed point of the system (1.14). For that purpose we choose a symmetric and positive definite real $k \times k - matrix$ B and define a function

$$V(x) = x^T B x , \; x \in \mathbb{R}^k.$$

If

$$x^T (A^T B A - B) x \le 0 \text{ for all } x \in \mathbb{R}^k,$$

then V is a *Lyapunov function* with respect to

$$f(x) = Ax , \; x \in \mathbb{R}^k,$$

on $G = \mathbb{R}^k$ which is *positive definite* with respect to $\{\Theta_k\}$.

Exercise 1.7.

a) Show with the aid of *Theorem 1.3* that $\{\Theta_k\}$ is stable with respect to f.
b) Show with the aid of *Theorem 1.3* that $\{\Theta_k\}$ is asymptotically stable with respect to f, if

$$x^T(A^T B A - B)x < 0 \text{ for all } x \in \mathbb{R}^k,\, x \neq \Theta_k.$$

One can even show that $\{\Theta_k\}$ is globally asymptotically stable with respect to f, i.e., $\{\Theta_k\}$ is stable with respect to f and

$$A^n x_0 \to \Theta_k \text{ as } n \to \infty \text{ for all } x_0 \in \mathbb{R}^k$$

which is equivalent to

$$A^n \to 0 = k \times k - zero\ matrix.$$

Conversely, let $\{\Theta_k\}$ be globally asymptotically stable with respect to f. Further let C be a symmetric and positive definite real $k \times k - matrix$ such that the series $\sum\limits_{k=0}^{\infty}(A^T)^k C A^k$ converges. If we define

$$\sum_{k=0}^{\infty}(A^T)^k C A^k,$$

then B is a symmetric and positive definite real $k \times k - matrix$ and it follows that

$$A^T B A - B = -C$$

which implies

$$x^T(A^T B A - B)x < 0 \text{ for all } x \in \mathbb{R}^k \text{ with } x \neq \Theta_k.$$

1.1.6 Applications

a) *Predator - Prey - Models*

The classical *predator - prey - model* as being developed and investigated by *Volterra* is based on a 2×2 - system of first order differential equations for the densities of the prey and predator population, respectively. This model has also been investigated in [13] with respect to stability of its equilibrium via a Lyapunov function.

Here we consider the discrete version of the model which is given by two difference equations of the form

$$x_{n+1} = (1+a)x_n - bx_n y_n \ ,$$
$$y_{n+1} = (1-c)y_n + dx_n y_n \ , \ n \in \mathbb{N}_0 \ , \tag{1.16}$$

with constant parameters $a > 0$, $0 < c < 1$, $b > 0$ and $d > 0$. The values x_n and y_n denote the density of the prey and predator population at time $t = n$, respectively. In this model it is assumed that the prey population grows exponentially in the absence of predators and that the predator population decays exponentially in the absence of prey. If we define

$$f(x,y) = \begin{pmatrix} f_1(x,y) \\ f_2(x,y) \end{pmatrix} \ , \ (x,y) \in \mathbb{R}^2 \ ,$$

with

$$f_1(x,y) = (1+a)x - bxy \text{ and } f_2(x,y) = (1-c)y + dxy \ ,$$

then (1.16) reads

$$\begin{pmatrix} x_{n+1} \\ y_{n+1} \end{pmatrix} = f(x_n, y_n) \ , \ n \in \mathbb{N}_0 \ , \tag{1.17}$$

and $f : \mathbb{R}^2 \to \mathbb{R}^2$ is a continuous mapping. The only fixed point $(x^*, y^*)^T \in \mathbb{R}^2$ with $x^* > 0$ and $y^* > 0$ is given by

$$x^* = \frac{c}{d} \text{ and } y^* = \frac{a}{b} \ .$$

For every $(x,y)^T \in \mathbb{R}^2$ the Jacobi matrix of f is given by

$$J_f(x,y) = \begin{pmatrix} 1+a-by & bx \\ dy & 1-c+dx \end{pmatrix}$$

which implies

$$J_f(x^*, y^*) = \begin{pmatrix} 1 & -\frac{bc}{d} \\ \frac{ad}{b} & 1 \end{pmatrix} .$$

The *eigenvalues* of $J_f(x^*, y^*)$ are given by

$$\lambda_{1,2} = 1 \pm i\sqrt{ac},$$

hence

$$|\lambda_{1,2}|^2 = 1 + ac > 1.$$

By the *Corollary of Theorem 1.5* the only fixed point $(x^*, y^*)^T$ of f in $\overset{\circ}{\mathbb{R}}^2_+$ is unstable. Next we assume that the prey population grows logistically in the absence of predators. Therefore we replace the first equation in (1.16) by

$$x_{n+1} = (1+a)x_n - ex_n^2 - bx_n y_n$$

with a constant parameter $e > 0$.

With this modification (1.16) can be written in the form (1.12) with

$$f_1(x_n, y_n) = (1 + a)x_n - ex_n^2 - bx_n y_n \text{ and}$$
$$f_2(x_n, y_n) = (1 - c)y_n + dx_n y_n , \; n \in \mathbb{N}_0 .$$

The only fixed point $(x^*, y^*)^T \in \overset{\circ}{\mathbb{R}}{}^2_+$ of f is then given by

$$x^* = \frac{c}{d} \text{ and } y^* = \frac{1}{b}\left(a - \frac{ce}{d}\right) ,$$

if

$$a > \frac{ce}{d} .$$

For every $(x, y)^T \in \mathbb{R}^2$ the Jacobi matrix of f is given by

$$J_f(x, y) = \begin{pmatrix} 1 + a - 2ex - by & -bx \\ dy & 1 - c + dx \end{pmatrix}$$

which implies

$$J_f(x^*, y^*) = \begin{pmatrix} 1 - e\frac{c}{d} & -\frac{bc}{d} \\ \frac{d}{b}\left(a - \frac{ce}{d}\right) & 1 \end{pmatrix}.$$

The eigenvalues of $J_f(x^*, y^*)$ are given by

$$\lambda_{1,2} = 1 - \frac{ec}{2d} \pm \sqrt{\left(\frac{ec}{2d}\right)^2 - c\left(a - \frac{ec}{d}\right)}.$$

We have to distinguish three cases:

1) $\left(\frac{ec}{2d}\right)^2 - c\left(a - \frac{ac}{d}\right) = 0$

   Then
   $$-1 < \lambda_{1,2} = 1 - \frac{ec}{2d} < 1 ,$$

   if and only if
   $$\frac{ec}{d} < 4.$$

2) $\left(\frac{ec}{2d}\right)^2 - c\left(a - \frac{ec}{d}\right) < 0.$

   Then
   $$|\lambda_{1,2}|^2 = 1 - \frac{ec}{d} + c\left(a - \frac{ec}{d}\right) < 1 ,$$

   if and only if
   $$a < \frac{ec}{d} + \frac{e}{d}.$$

3) $\left(\frac{ec}{2d}\right)^2 > c\left(a - \frac{ec}{d}\right)$.

Then

$$1 - \frac{ec}{2d} < \lambda_1 < 1 \text{ and } 1 - \frac{ec}{d} < \lambda_2 < 1.$$

Hence $-1 < \lambda_1 < 1$, if $-1 < 1 - \frac{ec}{2d} \Leftrightarrow \frac{ec}{d} < 4$

and $-1 < \lambda_2 < 1$, if $-1 < 1 - \frac{ec}{d} \Leftrightarrow \frac{ec}{d} < 2$.

From this we conclude that in all three cases

$$|\lambda_{1,2}| < 1 \text{ , if } a < \frac{2ec}{d} < 4.$$

Result. There exists exactly one fixed point $(x^*, y^*) \in \overset{\circ}{\mathbb{R}^2_+}$ of f and this fixed point is asymptotically stable, if

$$\frac{ce}{d} < a < \frac{2ec}{d} < 4.$$

Exercise 1.8. Show that the only fixed point of the system

$$x_{n+1} = (1 + a)x_n - bx_n y_n \ ,$$

$$y_{n+1} = (1 - c)y_n + d[(1 + a)x_n - bx_n y_n]y_n \ , \ n \in \mathbb{N}_0 \ ,$$

is asymptotically stable, if $a < 2$.

b) *A Discretization of the Volterra-Lotka Model*

This model is described by two differential equations of the form

$$\dot{x}(t) = (c_1 + c_{12}y(t))x(t) \ ,$$

$$\dot{y}(t) = (c_2 + c_{21}x(t))y(t)$$

where $x(t)$ and $y(t)$ denote the density of the prey and predator population, respectively, at the time t. The coefficients c_1, c_2, c_{12}, and c_{21} satisfy the conditions

$$c_1 > 0 \ , \ c_2 < 0 \ , \ c_{12} < 0 \ , \ c_{21} > 0 \ .$$

This leads to the fact that the above system of differential equations admits an equilibrium solution

$$x(t) = x^* = -\frac{c_2}{c_{21}} \ , \ y(t) = y* = -\frac{c_1}{c_{12}} \ , \ t \in \mathbb{R} \ .$$

We now discretize this system by introducing a time stepsize and replacing the derivatives $\dot{x}(t)$ and $\dot{y}(t)$ by the difference quotients

$$\frac{x(t+h)-x(t)}{h} \quad \text{and} \quad \frac{y(t+h)-y(t)}{h} \, ,$$

respectively. Then we obtain the system of difference equations

$$x(t+h) = x(t) + h(c_1 + c_{12}y(t))x(t) \, ,$$

$$(1.16')$$

$$y(t+h) = y(t) + h(c_2 + c_{21}x(t))y(t)$$

(see (1.16)). The above equilibrium solution of the system of differential equations turns out to be a fixed point solution of (1.16'), namely

$$x(t) = x^* \, , \quad y(t) = y^* \text{ for all } t = t_0 + kh \, , \quad k \in \mathbb{N}_0 \, ,$$

where $t_0 \in \mathbb{R}$ is chosen arbitrarily.
If we define

$$f(x,y) = \begin{pmatrix} f_1(x,y) \\ f_2(x,y) \end{pmatrix} \, , \quad (x,y) \in \mathbb{R}^2 \, ,$$

with

$$f_1(x,y) = x + h(c_1 + c_{12}y)x \quad \text{and} \quad f_2(x,y) = y + h(c_2 + c_{21}x)y \, ,$$

then (1.16') reads

$$\begin{pmatrix} x(t+h) \\ y(t+h) \end{pmatrix} = f(x(t), y(t)) \, . \tag{1.17'}$$

The Jacobi matrix of f at the fixed point (x^*, y^*) is given by

$$J_f(x^*, y^*) = \begin{pmatrix} 1 & -hc_{12}\frac{c_2}{c_{21}} \\ -hc_{21}\frac{c_1}{c_{12}} & 1 \end{pmatrix}$$

and has the eigenvalues

$$\lambda_{1,2} = 1 \pm ih\sqrt{|c_1| \cdot |c_2|}$$

which implies $|\lambda_{1,2}| > 1$.
By the Corollary of *Theorem 1.5* the fixed point (x^*, y^*) of f is unstable. Therefore we replace the system (1.16') by the system

$$x(t+h) = x(t) + h(c_1 + c_{12}y(t))x(t) \, ,$$

$$(1.16'')$$

$$y(t+h) = y(t) + h(c_2 + c_{21}x(t))y(t),$$

which has the same fixed point solution as (1.16′). The above vector function f has to be replaced by the function

$$f(x, y) = \begin{pmatrix} x + h(c_1 + c_{12}y)x \\ y + h(c_2 + c_{21}(x + h(c_1 + c_{12}y)x))y \end{pmatrix}, \quad (x, y) \in \mathbb{R}^2,$$

whose Jacobi matrix at the fixed point (x^*, y^*) is given by

$$J_f(x^*, y^*) = \begin{pmatrix} 1 & -hc_{12}\frac{c_2}{c_{21}} \\ -hc_{21}\frac{c_1}{c_{12}} & 1 + h^2 c_1 c_2 \end{pmatrix}.$$

The eigenvalues of $J_f(x^*, y^*)$ read

$$\lambda_{1,2} = 1 + \frac{h^2}{2}c_1 c_2 \pm \sqrt{\left(1 + \frac{h^2}{2}c_1 c_2\right) - 1}.$$

We distinguish three cases:
1) $\left(1 + \frac{h^2}{2}c_1 c_2\right)^2 = 1$.

Then it follows that

$$\lambda_1 = \lambda_2 = 1 + \frac{h^2}{2}c_1 c_2 = -1$$

which is equivalent to

$$|c_1 \cdot c_2| = \frac{4}{h^2} .$$

2) $\left(1 + \frac{h^2}{2}c_1 c_2\right)^2 < 1 \Leftrightarrow |c_1 \cdot c_2| < \frac{4}{h^2}$.

Then it follows that

$$\lambda_{1,2} = 1 + \frac{h^2}{2}c_1 c_2 \pm i\sqrt{1 - \left(1 + \frac{h^2}{2}c_1 c_2\right)^2}$$

which implies

$$\lambda_1 \neq \lambda_2 \quad \text{and} \quad |\lambda_1| = |\lambda_2| = 1 .$$

3) $\left(1 + \frac{h^2}{2}c_1 c_2\right)^2 > 1 \Leftrightarrow |c_1 \cdot c_2| > \frac{4}{h^2}$.

Then it follows that

$$\lambda_{1,2} = 1 + \frac{h^2}{2}c_1 c_2 \pm \sqrt{h^2 c_1 c_2 + \frac{h^4}{4}c_1^2 c_2^2}$$

$$= 1 + \frac{h^2}{2}c_1 c_2 \pm \frac{h^2}{2}|c_1 c_2|\sqrt{1 - \frac{4}{h^2|c_1 c_2|}}$$

which implies

$$\lambda_2 < \lambda_1 < 1 + \frac{h^2}{2}c_1c_2 + \frac{h^2}{2}|c_1 \cdot c_2| = 1$$

and

$$\lambda_2 > 1 + \frac{h^2}{2}c_1c_2 - \frac{h^2}{2}|c_1 \cdot c_2| = 1 - h^2|c_1 \cdot c_2| \ .$$

Further we have $\lambda_2 \geq -1$, if and only if

$$|c_1 \cdot c_2| \left(1 + \sqrt{1 - \frac{4}{h^2|c_1 \cdot c_2|}} \right) \leq \frac{4}{h^2} \quad (< |c_1 \cdot c_2|)$$

which is impossible.

Result:

If $|c_1 \cdot c_2| \leq \dfrac{4}{h^2}$, then $|\lambda_{1,2}| = 1$.

If $|c_1 \cdot c_2| > \dfrac{4}{h^2}$, then $\lambda_2 < \lambda_1 < 1$, but $\lambda_2 < -1$.

For the following we assume that

$$|c_1 \cdot c_2| < \frac{4}{h^2} \ .$$

Then it follows (as we have seen above) that

$$\lambda_1 \neq \lambda_2 \quad \text{and} \quad |\lambda_1| = |\lambda_2| = 1$$

which implies that the corresponding eigenvectors are linearly independent. From the remark following *Theorem 1.7* we therefore infer that the mapping $g : \mathbb{R}^2 \to \mathbb{R}^2$ defined by

$$g(x,y) = J_f(x^*, y^*) \begin{pmatrix} x \\ y \end{pmatrix} \quad , \quad x, \ y \in \mathbb{R} \ ,$$

has $\begin{pmatrix} 0 \\ 0 \end{pmatrix}$ as stable fixed point.

c) *Interacting Logistic Growth of Two Populations*

If we discretize the original model which is presented and investigated in [13] in terms of two first order differential equations, then we obtain two difference equations of the form

$$x_{n+1} = (1 + h\ a - h\ b\ x_n - h\ c\ y_n)x_n,$$
$$y_{n+1} = (1 + h\ d - h\ e\ x_n - h\ f\ y_n)y_n\ ,\ n \in \mathbb{N}_0 \tag{1.18}$$

with constant parameters $a, b, c, d, e, f > 0$ and step size $h > 0$. Again x_n and y_n denote the densities of the two populations at time $t = n$. Both populations grow logistically in the absence of the other population and the terms $(h\ c\ y_n x_n)$ and $(h\ e\ x_n y_n)$ describe the mutual interaction. If we define

$$f(x, y) = \begin{pmatrix} f_1(x, y) \\ f_2(x, y) \end{pmatrix}\ ,\ (x, y) \in \mathbb{R}^2,$$

with

$$f_1(x, y) = (1 + h\ a - h\ b\ x - h\ c\ y)x,$$
$$f_2(x, y) = (1 + h\ d - h\ e\ x - h\ f\ y)y,$$

then (1.18) reads

$$\begin{pmatrix} x_{n+1} \\ y_{n+1} \end{pmatrix} = f(x_n, y_n)\ ,\ n \in \mathbb{N}_0\ , \tag{1.19}$$

and $f : \mathbb{R}^2 \to \mathbb{R}^2$ is a continuous mapping.
The point $(x^*, y^*)^T \in \mathbb{R}^2$ is a fixed point of f with $x^* \neq 0$ and $y^* \neq 0$, if and only if

$$b\ x^* + c\ y^* = a\ ,$$
$$e\ x^* + f\ y^* = d\ .$$

Let us assume that

$$bf - ce > 0. \tag{1.20}$$

Then

$$x^* = \frac{a\ f - d\ c}{b\ f - e\ c} \quad \text{and} \quad y^* = \frac{a\ e - d\ b}{e\ c - b\ f}$$

and $x^* > 0$, $y^* > 0$, if and only if

$$\frac{c}{f} < \frac{a}{d} < \frac{b}{e}, \tag{1.21}$$

which implies (1.20).

For every $(x, y)^T \in \mathbb{R}^2$ the *Jacobi matrix* of f is given by

$$J_f(x, y) =$$

$$\begin{pmatrix} -h\,b\,x + 1 + h\,a - h\,b\,x - h\,c\,y & -h\,c\,x \\ -h\,e\,y & -h\,f\,y + 1 + h\,d - h\,e\,x - h\,f\,y \end{pmatrix}$$

which implies

$$J_f(x^*, y^*) = \begin{pmatrix} 1 - h\,b\,x^* & -h\,c\,x^* \\ -h\,e\,y^* & 1 - h\,f\,y^* \end{pmatrix}.$$

The *eigenvalues* of $J_f(x^*, y^*)$ are given by

$$\lambda_{1,2} = +1 + h \left(-\frac{bx^* + fy^*}{2} \pm \sqrt{\left(\frac{bx^* + fy^*}{2}\right)^2 - (bf - ec)x^*y^*} \right)$$

$$= +1 + h\mu_{1,2}.$$

From (1.21) which implies (1.20) and $x^* > 0$, $y^* > 0$ it follows that $\mathcal{R}e(\mu_{1,2}) < 0$. This implies $|\lambda_{1,2}| < 1$ (*Exercise*), if $h > 0$ is sufficiently small.

Result. If (1.21) is satisfied and $h > 0$ is sufficiently small, then there is exactly one fixed point $(x^*, y^*) \in \overset{\circ}{\mathbb{R}}{}^2_+$ of f and this fixed point is asymptotically stable.

Exercise 1.9. Show that the system

$$x_{n+1} = (1 + h\,a - h\,e\,x_n - h\,b\,y_n)x_n\,,$$

$$y_{n+1} = (1 - h\,c + h\,d\,x_n)y_n\,, \quad n \in \mathbb{N}_0\,,$$

with $a, b, c, d, e > 0$, $h > 0$ has exactly one fixed point $x^* > 0$, $y^* > 0$, if $a > \frac{c\,e}{d}$, which is asymptotically stable, if $h > 0$ is sufficiently small.

d) *An Emission Reduction Model*

In [25] a mathematical model for the reduction of carbon dioxide emission is investigated in form of a time discrete dynamical system which as un-controlled system is given by the following system of difference equations

$$E_i(t+1) = E_i(t) + \sum_{j=1}^{r} em_{ij} M_j(t) ,$$

$$M_i(t+1) = M_i(t) - \lambda_i M_i(t)(M_i^* - M_i(t))E_i(t) \qquad (1.22)$$

$$\text{for } i = 1, \dots, r \text{ and } t \in \mathbb{N}_0$$

where $E_i(t)$ denotes the amount of emission reduction and $M_i(t)$ the fi-nancial means spent by the $i - th$ *actor* at the time t, $\lambda_i > 0$ is a growth parameter and $M_i^* > 0$ an upper bound for $M_i(t)$ for $i = 1, \dots, r$ and $t \in \mathbb{N}_0$. For $t = 0$ we assume the system to be in the state E_{0i}, M_{0i}, $i = 1, \dots, r$ which leads to the initial conditions

$$E_i(0) = E_{0i} \text{ and } M_i(0) = M_{0i} \text{ for } 1, \dots, r.$$

If we define $x = (E^T, M^T)^T$, E, $M \in \mathbb{R}^r$, and functions $f_i : \mathbb{R}^n \to \mathbb{R}^n$, $i = 1, \dots, n = 2r$ by

$$f_i(x) = \qquad E_i + \sum_{j=1}^{r} em_{ij} M_j , \qquad i = 1, \dots, r , \qquad (1.23)$$

$$f_i(x) = M_i - \lambda_i M_i(M_i^* - M_i)E_i , \quad i = r+1, \dots, n , \qquad (1.24)$$

then we can write (1.22) in the form

$$x(t+1) = f(x(t)) , t \in \mathbb{N}_0 , \qquad (1.25)$$

where $f(x) = (f_1(x), \dots, f_n(x))^T$.

For every $\hat{x} = (\hat{E}^T, \Theta_r^T)^T$, $\hat{E} \in \mathbb{R}^r$, we have

$$\hat{x} = f(\hat{x}).$$

Let $\hat{x}$ be any such fixed point of f.

Then we replace the system (1.25) by the linear system

$$x(t+1) = J_f(\hat{x})x(t) \ , \ t \in \mathbb{N}_0 \ ,$$

$$(1.26)$$

where the *Jacobi matrix* $J_f(\hat{x})$ is given by

$$J_f(x) = \begin{pmatrix} I_r & C \\ O_r & D \end{pmatrix}$$

where I_r and O_r is the $r \times r - unit$ and $-zero\ matrix$, respectively, and

$$C = \begin{pmatrix} em_{11} & \cdots & em_{1r} \\ \vdots & \ddots & \vdots \\ em_{r1} & \cdots & em_{rr} \end{pmatrix}$$

and

$$D = \begin{pmatrix} d_{11} & & 0 \\ & \ddots & \\ 0 & & d_{rr} \end{pmatrix}$$

with

$$d_{ii} = 1 - \lambda_i M_i^* \hat{E}_i \ , \ i = 1, \ldots, r.$$

Let us assume that

$$d_{ii} \neq 0 \ , \ d_{ii} \neq 1 \ \text{ and } \ |d_{ii}| \leq 1 \text{ for all } i = 1, \ldots, r.$$

Then the matrix $J_f(\hat{x})$ is *non-singular*. Its *eigenvalues* are given by

$$\mu_i = 1 \text{ for } i = 1, \ldots, r \ , \ \mu_{1+r} = d_{ii} \text{ for } i = 1, \ldots, r,$$

hence

$$|\mu_i| \leq 1 \text{ for } i = 1, \ldots, 2r$$

and the corresponding *eigenvectors*

$$\begin{pmatrix} e_i \\ \Theta_r \end{pmatrix} \text{ for } j = 1, \ldots, r$$

and

$$\begin{pmatrix} -\frac{1}{1-d_{jj}} C e_j \\ e_j \end{pmatrix} \text{ for } j = 1, \ldots, r$$

(e_j = j-th unit vector) are linearly independent. *Theorem 1.7* therefore implies that the zero sequence $(x(t) = \Theta_n)_{t \in \mathbb{N}_0}$ which satisfies (1.26) is stable.

1.2 The Non-Autonomous Case

1.2.1 Definitions and Elementary Properties

Let $\overline{X}$ be a metric space with metric $d : \overline{X} \times \overline{X} \to \mathbb{R}_+$ and let $(f_n)_{n \in \mathbb{N}}$ be a sequence of continuous mappings $f_n : \overline{X} \to \overline{X}$, $n \in \mathbb{N}$. Then the pair $(\overline{X}, (f_n)_{n \in \mathbb{N}})$ is called a *non-autonomous* time-discrete dynamical system. The dynamics in this system is defined by the sequence $F = (F_n)_{n \in \mathbb{N}_0}$ of mappings $F_n : \overline{X} \to \overline{X}$ given by

$$F_n(x) = f_n \circ f_{n-1} \circ \ldots \circ f_1(x) \text{ for all } x \in \overline{X} \text{ and } n \in \mathbb{N}$$

and

$$F_0 = x \text{ for all } x \in \overline{X}.$$

If $f_n = f$ for all $n \in \mathbb{N}$ we obtain an *autonomous* time-discrete dynamical system $(\overline{X}, f)$ as being defined in *Section 1.1.1*.
For every $x \in \overline{X}$ we define an *orbit* starting with x by

$$\gamma_F(x) = \bigcup_{n \in \mathbb{N}_0} \{F_n(x)\} \tag{1.5'}$$

and the limit set of $\gamma_F(x)$ by

$$L_F(x) = \bigcap_{n \in \mathbb{N}_0} \overline{\bigcup_{m \geq n} \{F_m(x)\}} \tag{1.6'}$$

where $\overline{A}$ denotes the closure of $A \subseteq \overline{X}$.
Then we can prove

Proposition 1.1': *For every $x \in \overline{X}$ the limit set $L_F(x)$ consists of all accumulation points of the sequence $(F_n(x))_{n \in \mathbb{N}_0}$.*

The proof is the same as that of *Proposition 1.1* and is left as an exercise.

Now let $\overline{X}$ be compact. Then, for every $x \in \overline{X}$, the sequence $(F_n(x))_{n \in \mathbb{N}_0}$ has at least one accumulation point which implies, by *Proposition 1.1'*, that $L_F(x)$ is non-empty. By definition $L_F(x)$ is a closed subset of $\overline{X}$ and hence also compact for every $x \in \overline{X}$.
Further one can show that $L_F(x)$ is the smallest closed subset $S \subseteq \overline{X}$ with

$$\lim_{n \to \infty} \rho(F_n(x), S) = 0 \text{ with } \rho(y, S) = \min\{ d(y, z) \mid z \in \mathbb{S} \}. \tag{1.7'}$$

The proof is left as an exercise (see *Proposition 1.3*). In addition we can prove the following

Proposition 1.6. *Let* $\overline{X}$ *be compact and let* $(f_n)_{n \in \mathbb{N}}$ *be uniformly convergent to some mapping* $f_0 : \overline{X} \to \overline{X}$. *Then it follows that*

$$f_0(L_F(x)) = L_F(x) \text{ for all } x \in \overline{X}.$$

Proof.

1) At first we show that $f_0(L_F(x)) \subseteq L_F(x)$, $x \in \overline{X}$.
   Choose $x \in \overline{X}$ and $y \in L_F(x)$ arbitrarily. Since f_0 is continuous, for every $\varepsilon > 0$, there exists some $\delta = \delta(\varepsilon, y) > 0$ such that

   $$f_0(\tilde{x}) \in U_\varepsilon(f_0(y)) \text{ for all } \tilde{x} \in U_\delta(y).$$

   Uniform convergence of $(f_n)_{n \in \mathbb{N}}$ to f_0 implies the existence of some $n(\varepsilon) \in \mathbb{N}$ such that

   $$f_n(\tilde{x}) \in U_\varepsilon(f_0(\tilde{x})) \subseteq U_{2\varepsilon}(f_0(y)) \text{ for all } \tilde{x} \in U_\delta(y) \text{ and all } n \geq n(\varepsilon) .$$

   By *Proposition 1.1'* there is a subsequence $(F_{n_i}(x))_{i \in \mathbb{N}_0}$ of $(\mathbb{F}_n(x))_{n \in \mathbb{N}}$ with
   $$F_{n_i}(x) \in U_\delta(y) \text{ for all } i \in \mathbb{N}_0.$$

   This implies

   $$F_{n_i+1}(x) = f_{n_i+1} \circ F_{n_i}(x) \in U_{2\varepsilon}(f_0(y)) \text{ for all } n_i \geq n(\varepsilon).$$

   Since $\varepsilon > 0$ is chosen arbitrarily, it follows that

   $$F_{n_i+1}(x) \to f_0(y) \text{ and hence, by } \textit{Proposition (1.1')}, f_0(y) \in L_F(x).$$

2) Next we prove $L_F(x) \subseteq f_0(L_F(x))$, $x \in \overline{X}$.
   Choose $\hat{x} \in \overline{X}$ and $y \in L_F(\hat{x})$ arbitrarily. Then we have to show the existence of some $x \in L_F(\hat{x})$ such that $y = f_0(x)$.
   By *Proposition 1.1'* there is a subsequence $(F_{n_i}(\hat{x}))_{i \in \mathbb{N}_0}$ of $(F_n(\hat{x}))_{n \in \mathbb{N}_0}$ with $F_{n_i}(\hat{x}) \to y$ as $i \to \infty$. If we put

   $$x_i = F_{n_i-1}(\hat{x}) \text{ for all } i \in \mathbb{N}_0,$$

   then it follows that $f_{n_i}(x_i) \to y$ as $i \to \infty$. We can also assume that $x_i \to x$ for some $x \in L_F(\hat{x})$.

Then we have

$$
\begin{aligned}
d(f_0(x), y) \;\le\;& d(f_0(x), f_0(x_i)) + d(f_0(x_i), f_{n_i}(x_i)) + d(f_{n_i}(x_i), y) \\
\;\le\;& \underbrace{d(f_0(x), f_0(x_i))}_{\to\, 0} + \underbrace{\sup_{\tilde{x}\in\overline{X}}\; d(f_0(\tilde{x}), f_{n_i}(\tilde{x}))}_{\to\, 0} + \underbrace{d(f_{n_i}(x_i), y)}_{\to\, 0} \\
& \qquad\qquad\qquad\qquad \text{as } i \to \infty,
\end{aligned}
$$

hence $y = f_0(x)$.

$\square$

Next we give a more precise localization of the limit sets $L_F(x)$, $x \in \overline{X}$, with the aid of a *Lyapunov function* which is defined as follows:

Let $G \subseteq \overline{X}$, be nonempty. Then we say that $V : \overline{X} \to \mathbb{R}$ is a *Lyapunov function* with respect to $(f_n)_{n\in\mathbb{N}}$ on G, if
(1) V is continuous on $\overline{X}$,
(2) $V(f_n(x)) - V(x) \le 0$ for all $x \in G$ and all $n \in \mathbb{N}$ such that $f_n(x) \in G$.
For every $c \in \mathbb{R}$ we define

$$
V^{-1}(c) = \{x \in \overline{X} \mid V(x) = c\}.
$$

Then we can prove the following

Proposition 1.5': *Let V be a Lyapunov function with respect to $(f_n)_{n\in\mathbb{N}}$ on $G \subseteq \overline{X}$ where G is relatively compact. Let further $x_0 \in G$ be chosen such that $f_n(x_0) \in G$ for all $n \in \mathbb{N}$. Then there exists some $c \in \mathbb{R}$ such that*

$$
L_F(x_0) \subseteq V^{-1}(c)
$$

and $L_F(x_0)$ is nonempty.

Proof. Since G is relatively compact, it follows that $L_F(x_0)$ is nonempty. For every $n \in \mathbb{N}$ we put $x_n = F_n(x_0)$ which implies

$$
x_n \in G \quad \text{and} \quad V(x_{n+1}) \le V(x_n) \quad \text{for all } n \in \mathbb{N}.
$$

Since $V : \overline{X} \to \mathbb{R}$ is continuous, V is bounded from below on G which implies the existence of $c = \lim_{n\to\infty} V(x_n)$.
Now let $p \in L_F(x_0)$. Then, by *Proposition 1.1'*, there is a subsequence $(x_{n_i})_{i\in\mathbb{N}_0}$ of $(x_n)_{n\in\mathbb{N}_0}$ with $x_{n_i} \to p$ as $i \to \infty$.
This implies $V(p) = \lim_{i\to\infty} V(x_{n_i}) = c$, hence $p \in V^{-1}(c)$.

$\square$

1.2.2 Stability Based on Lyapunov's Method

Definition 1.8. *A relatively compact set $H \subseteq \overline{X}$ is called stable with respect to $(f_n)_{n \in \mathbb{N}}$, if for every relatively compact open set $U \subseteq \overline{X}$ with $U \supseteq \overline{H} =$ closure of H there exists an open set $W \subseteq \overline{X}$ with $\overline{H} \subseteq W \subseteq U$ such that*

$$F_n(W) \subseteq U \quad \text{for all} \quad n \in \mathbb{N}_0$$

where

$$F_n(W) = \{F_n(x) \mid x \in W\}.$$

Theorem 1.1': *Let $H \subseteq \overline{X}$ be relatively compact and such that for every relatively compact open set $U \subseteq \overline{X}$ with $U \supseteq \overline{H}$ there exists an open subset B_U of U with $B_U \supseteq \overline{H}$ and*

$$f_n(B_U) \subseteq U \quad \text{for all} \quad n \in \mathbb{N}.$$

Further let $G \subseteq \overline{X}$ be an open set with $G \supseteq \overline{H}$ such that there exists a Lyapunov function V with respect to $(f_n)_{n \in \mathbb{N}}$ on G which is positive definite with respect to $\overline{H}$, i.e.,

$$V(x) \geq 0 \quad for \; all \quad x \in G \quad and \quad (V(x) = 0 \Leftrightarrow x \in \overline{H}).$$

Then H is stable with respect to $(f_n)_{n \in \mathbb{N}}$.

Proof. Let $U \subseteq \overline{X}$ be an arbitrary relatively compact open set with $U \supseteq \overline{H}$. Then $U^* = U \cap G$ is also a relatively compact open set with $U^* \supseteq \overline{H}$ and there exists an open set $B_{U^*} \subseteq U^*$ with $B_{U^*} \supseteq \overline{H}$ and $f_n(B_{U^*}) \subseteq U^*$ for all $n \in \mathbb{N}$. Let us put

$$m = min\{V(x) \mid x \in U^* \backslash B_{U^*}\}.$$

Since $\overline{H} \cap (u^* \backslash B_{U^*})$ is empty, it follows that $m > 0$. If we define

$$W = \{x \in U^* \mid V(x) < m\},$$

then W is open and $\overline{H} \subseteq W \subseteq B_{U^*}$. Now let $x \in W$ be chosen arbitrarily. Then $x \in B_{U^*}$ and therefore $F_1(x) = f_1(x) \in U^*$. Further we have

$$V(F_1(x)) = V(f_1(x)) \leq V(x) < m,$$

hence $F_1(x) \in W \subseteq B_{U^*}$. This implies $F_2(x) = f_2(F_1(x)) \in U^*$ and $V(F_2(x)) \leq V(F_1(x)) < m$, hence $F_2(x) \in W$. By induction it therefore follows that

$$F_n(x) \in W \subseteq U^* \subseteq U \quad \text{for all} \quad n \in \mathbb{N}_0.$$

This shows that H is stable with respect to $(f_n)_{n \in \mathbb{N}}$.

$\square$

Definition 1.9. *A set* $H \subseteq \overline{X}$ *is called an attractor with respect to* $(f_n)_{n \in \mathbb{N}}$, *if there exists an open set* $U \subseteq \overline{X}$ *with* $U \supseteq \overline{H}$ *such that*

$$\lim_{n \to \infty} \rho(F_n(x), \overline{H}) = 0 \quad (\text{ in short}: \ F_n(x) \to \overline{H} \ ) \text{ for all } \ x \in U$$

where

$$\rho(y, \overline{H}) = \inf\{ \ d(y, z) \mid z \in \overline{H} \ \}.$$

If $H \subseteq \overline{X}$ *is stable and an attractor with respect to* $(f_n)_{n \in \mathbb{N}}$, *then* H *is called asymptotically stable with respect to* $(f_n)_{n \in \mathbb{N}}$.

Theorem 1.2': *Let* $H \subseteq \overline{X}$ *be such that there exists a relatively compact open set* $U \supseteq \overline{H}$ *with*

$$f_n(U) \subseteq U \quad \text{for all } \ n \in \mathbb{N}.$$

Further let $V : \overline{X} \to \mathbb{R}$ *be a Lyapunov function with respect to* $(f_n)_{n \in \mathbb{N}}$ *on u which is positive definite with respect to* $\overline{H}$. *Finally let*

$$\lim_{n \to \infty} V(F_n(x)) = 0 \quad \text{for all } \ x \in U.$$

Then H *is an attractor with respect to* $(f_n)_{n \in \mathbb{N}}$.

Proof. Let $x \in U$ be chosen arbitrarily. Since the sequence $(F_n(x))_{n \in \mathbb{N}_0}$ is contained in U, for every subsequence $(F_{n_i}(x))_{i \in \mathbb{N}_0}$ of $(F_n(x))_{n \in \mathbb{N}_0}$ there exists a subsequence $(F_{n_{i_j}}(x))_{j \in \mathbb{N}_0}$ and some $q \in \overline{U}$ with

$$\lim_{j \to \infty} F_{n_{i_j}}(x) = q.$$

This implies

$$\lim_{j \to \infty} V(F_{n_{i_j}}(x)) = V(q) = 0,$$

hence $q \in \overline{H}$, and therefore $F_{n_{i_j}}(x) \to \overline{H}$ as $j \to \infty$. From this it follows that $F_n(x) \to \overline{H}$ which shows that H is an attractor with respect to $(f_n)_{n \in \mathbb{N}}$.

$\square$

Corollary: *Under the assumptions of Theorem 1.1' and 1.2' it follows that* $H \subseteq \overline{X}$ *is asymptotically stable.*

Let us demonstrate *Theorem 1.1'* and *Theorem 1.2'* by an example which is a modification of *Example 1.8*. We choose $\overline{X} = \mathbb{R}^2$ and consider a sequence $(f_n)_{n \in \mathbb{N}}$ of mappings $f_n : \mathbb{R}^2 \times \mathbb{R}^2$ given by

$$f_n(x, y) = \left(\frac{a_n \, y}{1 + x^2} \ , \ \frac{b_n \, x}{1 + y^2} \right) \ , \ (x, y) \in \mathbb{R}^2 \ , \ n \in \mathbb{N} \ ,$$

where $(a_n)_{n\in\mathbb{N}}$ and $(b_n)_{n\in\mathbb{N}}$ are sequences of real numbers with

$$a_n^2 \leq 1 \quad \text{and} \quad b_n^2 \leq 1 \quad \text{for all} \ \ n \in \mathbb{N}. \tag{1.27}$$

Put $H = \{(0,0)\}$. If we choose

$$V(x,y) = x^2 + y^2 \ , \ (x,y) \in \mathbb{R}^2,$$

then V is a *Lyapunov function* on $G = \mathbb{R}^2$ with respect to $(f_n)_{n\in\mathbb{N}}$ which is positive definite with respect to $\overline{H} = \{(0,0)\}$, since

$$V(f_n(x,y)) - V(x,y) = \left(\frac{a_n^2}{(1+x^2)^2} - 1 \right) y^2 + \left(\frac{b_n^2}{(1+y^2)^2} - 1 \right) x^2$$

$$\leq (a_n^2 - 1)y^2 + (b_n^2 - 1)x^2 \leq 0 \quad \text{for all} \ \ (x,y) \in \mathbb{R}^2 \ \ \text{and} \ \ n \in \mathbb{N}.$$

Now let $U \subseteq \mathbb{R}^2$ be relatively compact and open with $(0,0) \in U$ be given. Then there exists some $r > 0$ such that

$$B_U = \{(x,y) \in \mathbb{R}^2 \mid x^2 + y^2 < r\} \subseteq U.$$

Further it follows that

$$f_{1n}(x,y)^2 + f_{2n}(x,y)^2 \leq x^2 + y^2 < r \quad \text{for all} \ \ (x,y) \in B_U \ ,$$

hence

$$f_n(B_U) \subseteq B_U \subseteq U \quad \text{for all} \ \ n \in \mathbb{N}.$$

Therefore all the assumptions of *Theorem 1.1'* hold true which implies that $\{(0,0)\}$ is stable with respect to $(f_n)_{n\in\mathbb{N}}$.
Next we sharpen (1.27) to

$$a_n^2 \leq \gamma < 1 \quad \text{and} \quad b_n^2 \leq \gamma < 1 \quad \text{for all} \ n \in \mathbb{N}.$$

Then it follows that

$$V(f_n(x,y)) \leq \gamma V(x,y) \quad \text{for all} \ \ n \in \mathbb{N} \ \ \text{and} \ \ (x,y) \in \mathbb{R}^2$$

which implies

$$V(F_n(x,y)) \leq \gamma^n V(x,y) \quad \text{for all} \ \ n \in \mathbb{N} \ \ \text{and} \ \ (x,y) \in \mathbb{R}^2$$

and in turn

$$\lim_{n\to\infty} V(F_n(x,y)) = 0 \quad \text{for all} \ \ (x,y) \in \mathbb{R}^2.$$

By *Theorem 1.2'*, $H = \{(0,0)\}$ is an attractor with respect to $(f_n)_{n\in\mathbb{N}}$ and by the *Corollary of Theorems 1.1' and 1.2'* $\{(0,0)\}$ is asymptotically stable with respect to $(f_n)_{n\in\mathbb{N}}$.

1.2.3 Linear Systems

We consider a normed linear space $(E, \| \cdot \|)$ and a sequence $(f_n)_{n \in \mathbb{N}}$ of mappings $f_n : E \to E$ which are given by

$$f_n(x) = A_n(x) + b_n \quad , \quad x \in E \quad , \quad n \in \mathbb{N} \, ,$$

where $(A_n)_{n \in \mathbb{N}}$ is a sequence of continuous linear mappings $A_n : E \to E$ and $(b_n)_{n \in \mathbb{N}}$ is a fixed sequence in E.

Then the pair $(E, (f_n)_{n \in \mathbb{N}})$ is a non-autonomous time-discrete dynamical system. The dynamics in this system is defined by the sequence $(F_n)_{n \in \mathbb{N}_0}$ of mappings $F_n : E \to E$ given by

$$F_n(x) = f_n \circ f_{n-1} \circ \cdots \circ f_1(x)$$

$$= A_n \circ A_{n-1} \circ \cdots \circ A_1(x) + \sum_{k=1}^{n} A_n \circ A_{n-1} \circ \cdots \circ A_{k+1}(b_k) \quad (1.28)$$

$$\text{for} \quad x \in E \quad \text{and} \quad n \in \mathbb{N}$$

where $A_n \circ A_{n+1} = \textit{identity mapping}$ for all $n \in \mathbb{N}$ and

$$F_0(x) = x \quad \text{for all} \quad x \in E. \quad (1.29)$$

In general there will be no common fixed point of all f_n, i.e., no point $x^* \in E$ with $x^* = f_n(x^*)$ for all $n \in \mathbb{N}$ which then would also be a common fixed point of all F_n. Therefore fixed point stability is not a reasonable concept in this case. We replace it by another concept of stability which has been introduced in *Section 1.1.5* already and whose definition will be repeated here.

Definition 1.10. *A sequence* $(x_n = F_n(x_0))_{n \in \mathbb{N}_0}$, $x_0 \in E$, *is called*

1. stable, *if for every* $\varepsilon > 0$ *and every* $N \in \mathbb{N}_0$ *there exists some* $\delta = \delta(\varepsilon, N) > 0$ *such that for every sequence* $(\bar{x}_n = F_n(\bar{x}_0))_{n \in \mathbb{N}_0}$, $\bar{x}_0 \in E$, *with* $\|\bar{x}_N - x_N\| < \delta$ *it follows that*

$$\|\bar{x}_n - x_n\| < \varepsilon \quad \text{for all} \quad n \geq N + 1.$$

2. attractive, *if for every* $N \in \mathbb{N}_0$ *there exists some* $\delta = \delta(N) > 0$ *such that for every sequence* $(\bar{x}_n = F_n(\bar{x}_0))_{n \in \mathbb{N}_0}$, $\bar{x}_0 \in E$, *with* $\|\bar{x}_N - x_N\| < \delta$ *it follows that*

$$\lim_{n \to \infty} \|\bar{x}_n - x_n\| = 0 \quad ;$$

3. asymptotically stable, *if* $(x_n = F_n(x_0))_{n \in \mathbb{N}_0}$, $x_0 \in E$, *is stable and attractive.*

As a first consequence of this definition we have

Lemma 1.1': *The following statements are equivalent:*
(1) All sequences $(x_n = F_n(x_0))_{n \in \mathbb{N}_0}$, $x_0 \in E$, are stable.
(2) One sequence $(x_n = F_n(x_0))_{n \in \mathbb{N}_0}$, $x_0 \in E$, is stable.
(3) The sequence $(x_n = A_n \circ A_{n-1} \circ \cdots \circ A_1(\Theta_E))_{n \in \mathbb{N}} = (x_n = \Theta_E)_{n \in \mathbb{N}_0}$ is stable.

The proof is the same as that of *Lemma 1.1*. We can also replace *stable* by *attractive* and hence by *asymptotically stable*. This *Lemma* leads to the following sufficient conditions for *stability* and *asymptotic stability*:

Theorem 1.8. *If*

$$\|A_n\| = \sup\{\|A_n(x)\| \mid \|x\| = 1\} \leq 1 \quad \text{for all} \ \ n \in \mathbb{N} \ , \tag{1.30}$$

then all sequences $(x_n = F_n(x_0))_{n \in \mathbb{N}_0}$, $x_0 \in E$, are stable.
If

$$\sup_{n \in \mathbb{N}} \|A_n\| < 1 \ , \tag{1.31}$$

then all sequences $(x_n = F_n(x_0))_{n \in \mathbb{N}_0}$, $x_0 \in E$, are asymptotically stable.

Proof. Let us first assume that (1.30) holds true. Let $\varepsilon > 0$, be chosen arbitrarily. Then we put $\delta = \varepsilon$ and conclude that for every sequence $(x_n = F_n(x_0))_{n \in \mathbb{N}_0}$, $x_0 \in E$, with $\|x_N\| < \delta$ for some $N \in \mathbb{N}_0$ it follows that

$$\|x_n - \Theta_E\| = \|A_n \circ A_{n-1} \circ \ldots \circ A_{N+1}(x_N)\|$$
$$\leq \|A_n\| \cdot \|A_{n-1}\| \cdots \|A_{N+1}\| \|x_N\| < \delta = \varepsilon \quad \text{for all} \ \ n \geq N+1 \ .$$

Hence the sequence $(x_n = \Theta_E)_{n \in \mathbb{N}_0}$ is stable and by *Lemma 1.1'* all sequences $(x_n = F_n(x_0))_{n \in \mathbb{N}_0}$, $x_0 \in E$ are stable.
Next we assume that (1.31) ($\Rightarrow$ (1.30)) holds true. Thus for every $N \in \mathbb{N}_0$ there exists some $\delta = \delta(N) > 0$ such that for every sequence $(x_n = F_n(x_0))_{n \in \mathbb{N}_0}$, $x_0 \in E$, with $\|x_N\| < \delta$ for some $N \in \mathbb{N}_0$ it follows that

$$\lim_{n \to \infty} \|x_n - \Theta_E\| = \lim_{n \to \infty} \|A_n \circ A_{n-1} \circ \cdots A_{N+1}(x_N)\|$$
$$\leq \lim_{n \to \infty} (\sup_{m \in \mathbb{N}} \|A_m\|)^{n-N-1} \|x_N\| = 0.$$

Thus the sequence $(x_n = \Theta_E)_{n \in \mathbb{N}_0}$ *is attractive*, hence *asymptotically stable* which implies that *all* sequences $(x_n = F_n(x_0))_{n \in \mathbb{N}_0}$, $x_0 \in E$, are *asymptotically stable*.
According to the above considerations it suffices to consider *homogeneous systems* with

$$f_n(x) = A_n(x) \ , \quad x \in E \ , \quad n \in \mathbb{N},$$

such that

$$F_n(x) = A_n \circ A_{n-1} \circ \cdots \circ A_1(x) \quad , \quad x \in E \quad , \quad n \in \mathbb{N}.$$

Then all sequences $(x_n = F_n(x_0))_{n\in\mathbb{N}_0}$, $x_0 \in E$, are *stable* or *asymptotically stable*, if and only if the sequence $(x_n = F_n(\Theta_E))_{n\in\mathbb{N}_0}$ is *stable* or *asymptotically stable*.

$\square$

This leads to

Theorem 1.9. *Assumption: All* A_n, $n \in \mathbb{N}$, *are invertible. Assertion: The sequence* $(A_n \circ A_{n-1} \circ \cdots \circ A_1(\Theta_E))_{n\in\mathbb{N}_0}$ *is stable, if and only if for every* $N \in \mathbb{N}_0$ *there exists a constant* $c_N > 0$ *such that*

$$\|A_n \circ A_{n-1} \circ \cdots \circ A_{N+1}\| \le c_N \quad \text{for all} \quad n \ge N+1. \tag{1.32}$$

The sequence $(A_n \circ A_{n-1} \circ \cdots \circ A_1(\Theta_E))_{n\in\mathbb{N}_0}$ *is asymptotically stable, if and only if for every* $N \in \mathbb{N}_0$

$$\lim_{n\to\infty} \|A_n \circ A_{n-1} \circ \cdots \circ A_{N+1}\| = 0. \tag{1.33}$$

Proof. If the sequence $(A_n \circ A_{n-1} \circ \cdots \circ A_1(\Theta_E))_{n\in\mathbb{N}_0}$ is stable, we choose an arbitrary $\varepsilon > 0$ and conclude that, for every $N \in \mathbb{N}_0$, there exists $\delta = \delta(\varepsilon, N) > 0$ such that for every sequence $(x_n = F_n(x_0))_{n\in\mathbb{N}_0}$, with $x_0 \in E$, it follows that

$$\|A_n \circ A_{n-1} \circ \cdots \circ A_{N+1}(x_N)\| < \varepsilon \quad \text{for all } n \ge N+1.$$

This implies

$$\|A_n \circ A_{n-1} \circ \cdots \circ A_{N+1}(x_N)\| < \varepsilon \quad \text{for all} \quad n \ge N+1 \quad \text{and} \quad \|x_N\| \le \frac{\delta}{2}$$

hence

$$\|A_n \circ A_{n-1} \circ \cdots \circ A_{N+1}\| =$$

$$\sup\{ \|A_n \circ A_{n-1} \circ \cdots \circ A_{N+1}(x_N)\| \mid \|x_N\| \le 1 \} \le \frac{2\varepsilon}{\delta} c_N$$

for all $n \ge N+1$ -here the Assumption that all A_n , $n \in \mathbb{N}$, are invertible is needed-. Conversely let (1.32) be true for every $N \in \mathbb{N}_0$. Then we choose $\varepsilon > 0$, $N \in \mathbb{N}$, define $\delta = \frac{\varepsilon}{c_N}$ and conclude that for every sequence $(x_N = F_n(x_0))_{n\in\mathbb{N}_0}$, $x_0 \in E$, with $\|x_N\| < \delta$ it follows that

$$\|x_n - \Theta_E\| = \|A_n \circ A_{n-1} \circ \cdots \circ A_{N+1}(x_N)\|$$
$$\le \|A_n \circ A_{n-1} \circ \cdots \circ A_{N+1}\| \, \|x_N\| < \varepsilon \quad \text{for all} \quad n \ge N+1.$$

Hence the sequence $(x_n = \Theta_E)_{n\in\mathbb{N}_0}$ is *stable*.

If the sequence $A_n \circ A_{n-1} \circ \cdots \circ A_1(\Theta_E)_{n\in\mathbb{N}_0}$, is *asymptotically stable*, hence *attractive*, then, for every $N \in \mathbb{N}_0$, there exists some $\delta = \delta(N) > 0$ such that for every sequence $(x_N = F_n(x_0))_{n\in\mathbb{N}_0}$, $x_0 \in E$, with $\|x_N\| < \delta$ it follows that

$$\lim_{n\to\infty} \|x_n - \Theta_E\| = 0, \quad \text{i.e., for every } \varepsilon > 0 \text{ there exists some } n(\varepsilon) \geq N+1$$

with $\|x_n - \Theta_E\| < \varepsilon$ for all $n \geq n(\varepsilon)$, hence

$$\|A_n \circ A_{n-1} \circ \cdots \circ A_{N+1}(x_N)\| < \varepsilon \text{ for all } n \geq n(\varepsilon) \text{ and } \|x_N\| \leq \frac{\delta}{2}.$$

This implies
$$\|A_n \circ A_{n-1} \circ \cdots \circ A_{N+1}\| =$$

$$\sup\{ \|A_n \circ A_{n-1} \circ \cdots \circ A_{N+1}(x_N)\| \mid \|x_N\| \leq 1 \} \leq \frac{2\,\varepsilon}{\delta}$$

(*here again the Assumption that all A_n , $n \in \mathbb{N}$, are invertible is needed*) for all $n \geq n(\varepsilon)$, hence

$$\lim_{n\to\infty} \|A_n \circ A_{n-1} \circ \cdots \circ A_{N+1}\| = 0.$$

Conversely let (1.33) be true for every $N \in \mathbb{N}_0$. Thus (1.32) is satisfied for every $N \in \mathbb{N}_0$ and therefore the sequence $(A_n \circ A_{n-1} \circ \cdots \circ A_1(\Theta_E)_{n\in\mathbb{N}_0})$ is *stable*. Further there exists $\delta = \delta(N) > 0$ such that for every sequence $(x_n = F_n(x_0))_{n\in\mathbb{N}_0}$, $x_0 \in E$, with $\|x_N\| < \delta$ it follows that

$$\lim_{n\to\infty} \|x_n - \Theta_E\| \leq \lim_{n\to\infty} \|A_n \circ A_{n-1} \circ \cdots \circ A_{N+1}\| \, \|x_N\| = 0.$$

Thus the sequence $(A_n \circ A_{n-1} \circ \cdots \circ A_1(\Theta_E))_{n\in\mathbb{N}_0}$ is *attractive*.

$\square$

Now we specialize to the case $E = \mathbb{R}^k$ equipped with any norm. Then we have
$$f_n(x) = A_n x + b_n , \quad x \in \mathbb{R}^k ,$$

where $(A_n)_{n\in\mathbb{N}}$ is a sequence of real $k \times k - matrices$ and $(b_n)_{n\in\mathbb{N}}$ a sequence of vectors $b_n \in \mathbb{R}^k$. This leads to

$$F_n(x) = f_n \circ f_{n-1} \circ \cdots \circ f_1(x)$$
$$= A_n A_{n-1} \cdots A_1 x + \sum_{j=1}^{n} A_n A_{n-1} \cdots A_{j+1} b_j$$

for $x \in \mathbb{R}^k$ and $n \in \mathbb{N}$ where $A_n A_{n-1} \dots A_{j+1} b_j = b_n$ for $j = n$.

Let us assume that, for some $p \in \mathbb{N}$,

$$A_{n+p} = A_n \text{ and } b_{n+p} = b_n$$

for all $n \in \mathbb{N}$.

The question then is whether there is a sequence $(x_n)_{n \in \mathbb{N}_0}$ with

$$x_n = A_n x_{n-1} + b_n \quad \text{for all} \quad n \in \mathbb{N} \tag{1.34}$$

such that

$$x_{n+p} = x_n \quad \text{for all } n \in \mathbb{N}_0 \tag{1.35}$$

which implies $x_p = x_0$.

Conversely let this be the case. Then

$$
\begin{aligned}
x_{1+p} &= A_{1+p} x_p + b_{1+p} & = A_1 x_0 + b_1 & = x_1 \\
x_{2+p} &= A_{2+p} x_{1+p} + b_{1+p} & = A_2 x_1 + b_2 & = x_2 \\
&\;\;\vdots & \vdots \qquad & \;\;\vdots \\
x_{n+p} &= A_{n+p} x_{n-1+p} + b_{n+p} = A_n x_{n-1} + b_n = x_n \quad,
\end{aligned}
$$

hence

$$x_{x+p} = x_n \quad \text{for all} \quad n \in \mathbb{N}_0.$$

Result. A sequence $(x_n)_{n \in \mathbb{N}_0}$ with (1.34) satisfies (1.35), if and only if $x_p = x_0$ which is equivalent to $F_p(x_0) = x_0$, i.e.,

$$x_0 = A_p A_{p-1} \cdots A_1 x_0 + \sum_{j=1}^{p} A_p A_{p-1} \cdots A_{j+1} b_j.$$

This implies that there exists exactly one sequence $(x_n)_{n \in \mathbb{N}_0}$ with (1.34) and (1.35), if and only if the matrix $I - A_p A_{p-1} \ldots A_1$, $I = k \times k$-unit matrix, is non-singular. The first vector x_0 is then given by

$$x_0 = (I - A_p A_{p-1} \ldots A_1)^{-1} \sum_{j=1}^{p} A_p A_{p-1} \ldots A_{j+1} A_j.$$

In the homogeneous case where

$$b_n = \Theta_k \text{ for all } n \in \mathbb{N}$$

a non-zero sequence $(x_n)_{n \in \mathbb{N}_0}$ with (1.34) and (1.35) can only exist, if the matrix $I - A_p A_{p-1} \ldots A_1$ is singular,

i.e.

$$\det(I - A_p A_{p-1} \ldots A_1) = 0 \,,$$

This again is equivalent to the fact that $\lambda = 1$ is an *eigenvalue* of $A_p A_{p-1} \ldots A_1$.

1.2.4 Application to a Model for the Process of Hemo-Dialysis

In order to describe the temporal development of the concentration of some poison like urea in the body of a person suffering from a renal disease and having to be attached to an artificial kidney a mathematical model has been proposed in [12] which can be described as follows. The human body is divided into two compartments being termed as cellular part Z and extracellular part E of volume V_Z and V_E, respectively.

The two compartments are separated by cell membranes having the permeability C_Z [ml/min]. Let t[min] denote the time and let $K_Z(t)$ [mg/min] and $K_E(t)$ [mg/min] be the concentration of the poison at the time t in Z, and E, respectively. We consider some time interval $[0, T]$, $T > 0$, and assume that the patient is attached to the artificial kidney during the subinterval $[0, t_d]$ for some $t_d \in (0, T]$. We further assume that the generation rate of the poison in Z and E is L_1 [mg/min] and L_2 [mg/min], respectively. Then the temporal development of $K_Z = K_Z(t)$ and $K_E = K_E(t)$ in $[0, \infty)$ can be described by the following system of linear differential equations

$$
\begin{aligned}
V_Z \dot{K}_Z(t) &= -C_Z(K_Z(t) - K_E(t)) + L_1 \ , \\
V_E \dot{K}_E(t) &= C_Z(K_Z(t) - K_E(t)) - C(t) K_E + L_2
\end{aligned}
\tag{1.36}
$$

where

$$
C(t) = \begin{cases} C & \text{for } 0 \le t < t_d, \\ 0 & \text{for } t_d \le t < T. \end{cases}
\tag{1.37}
$$

$$
C(t + T) = C(t) \quad \text{for all } t \ge 0
\tag{1.38}
$$

and C [ml/min] is the permeability of the membrane of the artificial kidney.

By (1.38) the periodicity of the process of dialysis is expressed. The main result in [12] is the proof of the existence of exactly one pair $(K_Z(t) , K_E(t))$ of positive and T-periodic solutions of (1.36) which intuitively is to be expected. For the numerical computation of these solutions Euler's polygon method can be used. For this purpose we choose a time stepsize $\Delta t > 0$ such that

$$
t_d = K \cdot \Delta t \ , \quad T = N \cdot \Delta t
$$

$$
\text{for } K , N \in \mathbb{N} , \ 2 < K < N
$$

and replace (1.36) by the difference equations

$$V_Z(K_Z(t + \Delta t) - K_Z(t)) = -C_Z \Delta t (K_Z(t) - K_E(t)) + L_1 \Delta t \; ,$$

$$(1.39)$$

$$V_E(K_E(t + \Delta t) - K_E(t)) = C_Z \Delta t (K_Z(t) - K_E(t)) - C(t) \Delta t K_E(t) + L_2 \Delta t.$$

If we define, for every $n \in \mathbb{N}_0$,

$$x_n = \begin{pmatrix} K_Z(n \cdot \Delta t) \\ K_E(n \cdot \Delta t) \end{pmatrix} ,$$

$$A_{n+1} = \begin{pmatrix} 1 - C_Z \Delta t / V_Z & C_Z \Delta t / V_Z \\ C_Z \Delta t / V_E & 1 - (C(n \cdot \Delta t) + C_Z) \Delta t / V_E \end{pmatrix}$$

and

$$b_{n+1} = \begin{pmatrix} L_1 \Delta t / V_Z \\ L_2 \Delta t / V_E \end{pmatrix} = b \; ,$$

then (1.39) can be rewritten in the form

$$x_{n+1} = A_{n+1} \, x_n + b_{n+1} \; , \; n \in \mathbb{N}_0 \; , \qquad (1.40)$$

and we have

$$A_{n+N} = A_n \quad \text{and} \quad b_n = b \quad \text{for all} \quad n \in \mathbb{N}.$$

In order to show the existence and uniqueness of a periodic solution of (1.39) we assume that

$$\Delta t < \min(V_Z / C_Z) \; , \; V_E / (C + C_Z)). \qquad (1.41)$$

We have to show that the matrix $I - A_N A_{N-1} \ldots A_1$ with $I = \begin{pmatrix} 1 & 0 \\ 0 & 1 \end{pmatrix}$ is non-singular.
Let us put

$$c_1 = C_Z \Delta t / V_Z \quad c_2 = C_Z \Delta t / V_E.$$

Then, for all $n = K, \ldots, N - 1$, we obtain

$$A_n = \begin{pmatrix} 1 - c_1 & c_1 \\ c_2 & 1 - c_2 \end{pmatrix}$$

and hence

$$A_n \begin{pmatrix} 1 \\ 1 \end{pmatrix} = \begin{pmatrix} 1 \\ 1 \end{pmatrix} .$$

If we put
$$c_3 = C\Delta t/V_E \; ,$$

then we obtain
$$A_n = \begin{pmatrix} 1 - c_1 & c_1 \\ c_2 & 1 - c_2 - c_3 \end{pmatrix} \quad \text{for} \quad n = 1, \ldots, K - 1.$$

This implies
$$A_n \begin{pmatrix} 1 \\ 1 \end{pmatrix} = \begin{pmatrix} 1 \\ 1 - c_3 \end{pmatrix} \le \begin{pmatrix} 1 \\ 1 \end{pmatrix}, \quad \text{for} \quad n = 1, \ldots, K - 1$$

and hence
$$A_{K-2}A_{K-3} \ldots A_1 \begin{pmatrix} 1 \\ 1 \end{pmatrix} \le \begin{pmatrix} 1 \\ 1 - c_3 \end{pmatrix}.$$

Therefore
$$A_{K-1}A_{K-2} \ldots A_1 \begin{pmatrix} 1 \\ 1 \end{pmatrix} \le \begin{pmatrix} 1 \\ 1 - c_3 - c_3(1 - c_2 - c_3) \end{pmatrix} < \begin{pmatrix} 1 \\ 1 \end{pmatrix}$$

and finally
$$A_N A_{N-1} \ldots A_1 \begin{pmatrix} 1 \\ 1 \end{pmatrix} < \begin{pmatrix} 1 \\ 1 \end{pmatrix}.$$

Since the spectral radius $\varrho(B)$ of every quadratic matrix B with positive elements satisfies the inequality

$$\varrho(B) \le \max(By)_i/y_i$$

for all $y = (y)_i$ with $y_i > 0$ for all i (see [5]), it follows that

$$\varrho(A_N A_{N-1} \ldots A_1) < 1$$

which implies that $I - A_N A_{N-1} \ldots A_1$ is non-singular.
Hence there exists exactly one N-periodic solution of (1.40).
This is also positive. In order to see that we have to show that x_0 which is given by

$$x_0 = (I - A_N A_{N-1} \ldots A_1)^{-1} \sum_{j=1}^{N} A_N A_{N-1} \ldots A_{j+1} b_j$$

is positive.

First of all we observe that

$$\sum_{j=1}^{N} A_N A_{N-1} \ldots A_{j+1} b_j$$

is positive. If we define $B = A_N A_{N-1} \ldots A_1$, then B has positive elements and therefore

$$(I - B)^{-1} = \sum_{j=0}^{\infty} B^j$$

has positive elements which implies that x_0 is positive and in turn that all x_n, $n \in \mathbb{N}_0$, are positive.

2

Controlled Systems

2.1 The Autonomous Case

2.1.1 The Problem of Fixed Point Controllability

We begin with a system of difference equations of the form

$$x(t + 1) = g(x(t), u(t)) , \ t \in \mathbb{N}_0 \qquad (2.1)$$

where $g : \mathbb{R}^n \times \mathbb{R}^m \to \mathbb{R}^n$ is a continuous mapping.
The functions $x : \mathbb{N}_0 \to \mathbb{R}^n$ and $u : \mathbb{N}_0 \to \mathbb{R}^m$ are considered as state and control functions, respectively. For every control function $u : \mathbb{N}_0 \to \mathbb{R}^m$ and every vector $x_0 \in \mathbb{R}^n$ there exists exactly one state function $x : \mathbb{N}_0 \to \mathbb{R}^n$ which satisfies (2.1) and

$$x(0) = x_0. \qquad (2.2)$$

If we fix the control function $u : \mathbb{N}_0 \to \mathbb{R}^m$ and define

$$f_{t+1}(x) = g(x, u(t)) , \ x \in \mathbb{R}^n , \ t \in \mathbb{N}_0 , \qquad (2.3)$$

then, for every $t \in \mathbb{N}$, $f_t : \mathbb{R}^n \to \mathbb{R}^n$ is a continuous mapping and $(\mathbb{R}^n, (f_t)_{t \in \mathbb{N}})$ is a *non-autonomous* time-discrete dynamical system which is controlled by the function $u : \mathbb{N}_0 \to \mathbb{R}^m$. If

$$u(t) = \Theta_m = \text{ zero vector of } \mathbb{R}^m \text{ for all } t \in \mathbb{N}_0 ,$$

then the system (2.1) is called *uncontrolled*. Let us assume that the uncontrolled system (2.1) admits fixed points $\hat{x} \in \mathbb{R}^n$ which then solve the equation

$$\hat{x} = g(\hat{x}, \Theta_m). \qquad (2.4)$$

Now let $\Omega \subseteq \mathbb{R}^m$ be a subset with $\Theta_m \in \Omega$. Then we define the set of admissible control functions by

$$U = \{u : \mathbb{N}_0 \to \mathbb{R}^m \mid u(t) \in \Omega \text{ for all } t \in \mathbb{N}_0\}. \qquad (2.5)$$

After these preparations we are in the position to formulate the

Problem of Fixed Point Controllability

Given a fixed point $\hat{x} \in \mathbb{R}^n$ of the system

$$x(t + 1) = g(x(t), \Theta_m)) \; , \; t \in \mathbb{N}_0 \tag{2.6}$$

i.e., a solution $\hat{x}$ of equation (2.4) and an initial state $x_0 \in \mathbb{R}^n$ find some $N \in \mathbb{N}_0$ and a control function $u \in U$ with

$$u(t) = \Theta_m \quad \text{for all} \; t \geq N \tag{2.7}$$

such that the solution $x : \mathbb{N}_0 \to \mathbb{R}^n$ of (2.1), (2.2) satisfies the end condition

$$x(N) = \hat{x} \tag{2.8}$$

(which implies $x(t) = \hat{x}$ for all $t \geq N$.) From (2.1) and (2.2) it follows that

$$x(N) = \underbrace{g(g(\ldots(g(x_0, u(0)), u(1)), \ldots), u(N-1))}_{N-times}$$

$$\tag{2.9}$$

$$= G^N(x_0, u(0), \ldots, u(N-1)).$$

Let $N \in \mathbb{N}$, be given. If $u(0), \ldots, u(N-1) \in \Omega$ are solutions of the system

$$G^N(x_0, u(0), \ldots, u(N-1)) = \hat{x} \tag{2.10}$$

and one defines

$$u(t) = \Theta_m \quad \text{for all} \; t \geq N \; ,$$

then one obtains a control function $u : \mathbb{N}_0 \to \mathbb{R}^m$ which solves the problem of fixed point controllability.
From the definition (2.9) it follows that

$$G^N(x_0, u(0), \ldots, u(N-1)) = G^1(G^{N-1}(x_0, u(0), \ldots, u(N-2)), u(N-1)).$$

Conversely now let us assume that we are given a sequence $(G^N)_{N \in \mathbb{N}}$ of vector functions $G^N : \mathbb{R}^n \times \mathbb{R}^{m \cdot N} \to \mathbb{R}^n$ with this property.
Then we define, for every $t \in \mathbb{N}$,

$$x(t) = G^t(x_0, u(0), \ldots, u(t-1))$$
$$\text{for} \; x_0 \in \mathbb{R}^n \; \text{and} \; u(s) \in \mathbb{R}^m \; \text{for} \; s = 0, \ldots, t-1$$

and conclude

$$x(t) = G^1(G^{t-1}(x_0, u(0), \ldots, u(t-2)), u(t-1))$$
$$= G^1(x(t-1), u(t-1))$$
$$\text{for all } t \in \mathbb{N},$$

if we define $G^0(x_0) = x_0$. Now let $S(\hat{x}) \subseteq \mathbb{R}^n$ be the set of all vectors $x_0 \in \mathbb{R}^n$ such that there exists a time $N \in \mathbb{N}$ and a solution $(u(0)^T, \ldots, u(N-1)^T)^T \in \Omega^N$ of the system (2.10). Obviously, $\hat{x} \in S(\hat{x})$ (choose $N = 1$ and $u(0) = \Theta_n$). A first simple sufficient condition for the solvability of the problem of fixed point controllability is then given by

Proposition 2.1. *Let $\hat{x} \in S(\hat{x})$ be an interior point of $S(\hat{x})$ and let $\hat{x}$ be an attractor of the uncontrolled system (2.6), i.e., there exists an open set $U \subseteq \mathbb{R}^n$ with $\hat{x} \in U$ such that $\lim\limits_{t\to\infty} x(t) = \hat{x}$ where $x : \mathbb{N}_0 \to \mathbb{R}^n$ is a solution of (2.6) with (2.2) for any $x_0 \in U$. Then it follows that $S(\hat{x}) \supseteq U$ which implies that for every choice of $x_0 \in U$ the problem of fixed point controllability has a solution.*

Proof. Since $\hat{x} \in S(\hat{x})$ is an interior point of $S(\hat{x})$, there is an open neighborhood $W(\hat{x}) \subseteq \mathbb{R}^n$ of $\hat{x}$ with $W(\hat{x}) \subseteq S(\hat{x})$.
Now let $x_0 \in U$ be chosen arbitrarily. Then there is some $N_1 \in \mathbb{N}$ with $x(N_1) \in W(\hat{x})$ where $x : \mathbb{N}_0 \to \mathbb{R}^n$ is a solution of (2.6), (2.2). This implies the existence of some $N_2 \in \mathbb{N}$ and a solution $(u(0)^T, \ldots, u(N_2-1)^T)^T \in \Omega^{N_2}$ with

$$G^{N_2}(x(N_1), u(0), \ldots, u(N_2-1)) = \hat{x}.$$

If we define

$$u^*(t) = \begin{cases} \Theta_m & \text{for } t = 0, \ldots, N_1 - 1, \\ u(t - N_1) & \text{for } t = N_1, \ldots, N_1 + N_2 - 1, \end{cases}$$

then it follows that

$$G^N(x_0, u^*(0), \ldots, u^*(N-1)) = \hat{x}$$

where $N = N_1 + N_2$, i.e., $x_0 \in S(\hat{x})$ which completes the proof.

$\square$

The essential assumption in *Proposition 2.1* is that the fixed point $\hat{x}$ of the uncontrolled system (2.6) is an interior point of the controllable set $S(\hat{x})$.

In order to find sufficient conditions for that we assume that Ω is open and $g \in C^1(\mathbb{R}^n \times \mathbb{R}^m)$ which implies $G^N \in C^1(\mathbb{R}^n \times \mathbb{R}^{m \cdot N})$ for every $N \in \mathbb{N}$ and

$$
\begin{aligned}
G_x^N(x, u(0), \ldots, u(N-1)) = & \\
g_x(G^{N-1}(x, u(0), \ldots, u(N-2)), u(N-1)) & \times \\
\vdots & \\
g_x(x, u(0)),
\end{aligned}
$$

and

$$
\begin{aligned}
G_{u(k)}^N(x, u(0), \ldots, u(N-1)) = & \\
g_x(G^{N-1}(x, u(0), \ldots, u(N-2)), u(N-1)) & \times \\
g_x(G^{N-2}(x, u(0), \ldots, u(N-3)), u(N-2)) & \times \\
\vdots & \\
g_x(G^{k+1}(x, u(0), \ldots, u(k)), u(k+1)) & \times \\
g_{u(k)}(G^k(x, u(0), \ldots, u(k-1)), u(k)) &
\end{aligned}
$$

for $k = 0, \ldots, N-1$.

Let us assume that $g_x(\hat{x}, \Theta_m)$ is non-singular. Then it follows, for all $N \in \mathbb{N}$, that $G_x^N(\hat{x}, \Theta_m, \ldots, \Theta_m)$ is also non-singular, since

$$
G_x^N(\hat{x}, \Theta_m, \ldots, \Theta_m) = g_x(\hat{x}, \Theta_m)^N.
$$

Since

$$
\hat{x} = G^N(\hat{x}, \Theta_m, \ldots, \Theta_m) ,
$$

there exists, by the implicit function theorem, an open set $V \subseteq \Omega^N$ with $\Theta_m^N \in V$ and a function $h : V \to \mathbb{R}^n$ with $h \in C^1(V)$ such that

$$
h(\Theta_m^N) = \hat{x} \quad \text{and} \quad G^N(h(u(0), \ldots, u(N-1)), u(0), \ldots, u(N-1)) = \hat{x}
$$

for all $(u(0), \ldots, u(N-1)) \in V$.
Moreover,

$$
\begin{aligned}
h_{u(k)}(\Theta_m^N) &= -G_x^N(\hat{x}, \Theta_m, \ldots, \Theta_m)^{-1} G_{u(k)}(\hat{x}, \Theta_m, \ldots, \Theta_m) \\
&= -g_x(\hat{x}, \Theta_m)^{-N} g_x(\hat{x}, \Theta_m)^{N-k} g_{u(k)}(\hat{x}, \Theta_m) \\
&= -g_x(\hat{x}, \Theta_m)^{-k} g_{u(k)}(\hat{x}, \Theta_m)
\end{aligned}
$$

for $k = 0, \ldots, N-1$.

Result. If $g_x(\hat{x}, \Theta_M)$ is non-singular, then, for every $N \in \mathbb{N}$, there is an open set $V_N \subseteq \Omega^N$ with $\Theta_m^N \in V_N$ and a function $h_N : V_N \to \mathbb{R}^n$ with $h_N \in C^1(V_N)$ such that

$$h_N(u(0), \dots, u(N-1)) \in S(\hat{x})$$

for all $(u(0), \dots, u(N-1)) \in V_N$.

Next we assume that, for some $N \in \mathbb{N}$,

$$rank(h_{u(0)}(\Theta_m^N) \mid \dots \mid h_{u(N-1)}(\Theta_m^N)) = n.$$

Then there are n columns in the $n \times m \cdot N$-matrix

$$(h_{u(0)}(\Theta_m^N) \mid \dots \mid h_{u(N-1)}(\Theta_m^N))$$

which are linearly independent.

Now let E be the n-dimensional subspace of $\mathbb{R}^{m \cdot N}$ consisting of all vectors whose components vanish that do not correspond to the above linearly independent columns.

If we put $U = E \cap V_N$, then U is an open subset of E and the restriction of h to U is a C^1-function on U whose Jacobi matrix at Θ_m^N consists of the above linearly independent columns of $(h_{u(0)}(\Theta_m^N) \mid \dots \mid h_{u(N-1)}(\Theta_m^N))$ and is therefore invertible. By the inverse function theorem there exist open sets $\tilde{U} \subseteq E \cap V_N$ and $\tilde{V} \subseteq \mathbb{R}^n$ with $\Theta_m^N \in \tilde{U}$ and $\hat{x} \in \tilde{V}$ such that h is homeomorphic on $\tilde{U}$ and $h(\tilde{U}) = \tilde{V}$.

This implies that $\hat{x}$ is an interior point of $S(\hat{x})$.

Let us demonstrate this result by the *predator prey model* that was investigated with respect to asymptotic stability in *Section 1.1.6*. We consider this model as a controlled system of the form

$$x_1(t+1) = x_1(t) + ax_1(t) - ex_1(t)^2 - bx_1(t)x_2(t) - x_1(t)u_1(t),$$
$$x_2(t+1) = x_2(t) - cx_2(t) + dx_1(t)x_2(t) - x_2(t)u_2(t) , \ t \in \mathbb{N}_0.$$

The mapping $g : \mathbb{R}^2 \times \mathbb{R}^2 \to \mathbb{R}^2$ in (2.1) is therefore given by

$$\begin{pmatrix} (1+a)\, x_1 - e\, x_1^2 - b\, x_1x_2 - x_1\, u_1 \\ (1-c)\, x_2 + d\, x_1x_2 - x_2\, u_2 \end{pmatrix} , \ x, \ u \in \mathbb{R}^2.$$

We have seen in *Section 1.1.6* that

$$\hat{x} = (\frac{c}{d}, \frac{1}{b}(a - \frac{c\,e}{d}))^T$$

is the only fixed point of the uncontrolled system (with $u_1 = u_2 = 0$) in $\overset{\circ}{\mathbb{R}}_+^2 \times \overset{\circ}{\mathbb{R}}_+^2$, if $a > \frac{c\,e}{d}$.

One calculates

$$g_x(\hat{x}, \Theta_2) = \begin{pmatrix} 1 - e^{\frac{c}{d}} & -\frac{b}{d}\frac{c}{d} \\ \frac{d}{b}(a - \frac{c\,e}{d}) & 1 \end{pmatrix}$$

which implies that $g_x(\hat{x}, \Theta_2)$ is non-singular, if and only if

$$1 - e\frac{c}{d} + c(a - \frac{c\,e}{d}) \neq 0 \qquad (*)$$

Further we obtain

$$g_x(\hat{x}, \Theta_2) = \begin{pmatrix} -\frac{c}{d} & 0 \\ 0 & -\frac{1}{b}(a - \frac{c\,e}{d}) \end{pmatrix}$$

which implies

$$rank(h_{u(0)}(\Theta_2)) = rank(-g_{u(0)}(\hat{x}, \Theta_2)) = 2.$$

Hence $\hat{x}$ is an interior point of $S(\hat{x})$, if $(*)$ is satisfied.
This example is a special case of the following situation:
Let

$$g(x, u) = f(x) + F(x)u = f(x) + \sum_{i=1}^{m} f_i(x)u_i ,$$

$$x \in \mathbb{R}^n , \; u_1, \ldots, u_m \in \mathbb{R} ,$$

where $f, \; f_i \in C^1(\mathbb{R}^n), \; i = 1, \ldots, m$.
Let $\hat{x} \in \mathbb{R}^n$ be a fixed point of f, i.e.,

$$\hat{x} = f(\hat{x}) = g(\hat{x}, \Theta_m).$$

Then

$$g_x(\hat{x}, \Theta_m) = f_x(\hat{x}) \; \text{ and } \; g_u(\hat{x}, \Theta_m) = F(\hat{x})$$

and hence

$$((h_{u(0)}(\Theta_m^N) \mid \ldots \mid h_{u(N-1)}(\Theta_m^N))$$
$$= -(F(\hat{x}) \mid f_x(\hat{x})^{-1}F(\hat{x}) \mid \ldots \mid f_x(\hat{x})^{-N+1}F(\hat{x})).$$

In the example we have $m = n = 2$ and the 2×2-matrix $F(\hat{x})$ is non-singular.
Next we come back to the solution of (2.10) which we replace by an optimization problem. For this purpose we define a cost functional $\varphi : \mathbb{R}^{m \cdot N} \to \mathbb{R}$ by putting

$$\varphi(u(0), \ldots, u(N-1)) = \|G^N(x_0, u(0), \ldots, u(N-1)) - \hat{x}\|_2^2$$

for $(u(0), \ldots, u(N-1)) \in \mathbb{R}^{m \cdot N}$ $\quad (\|\cdot\|_2 = \text{Euclidean norm in } \mathbb{R}^n)$

and try to find $(u(0), \ldots, u(N-1)) \in \Omega^N$ such that $\varphi(u(0), \ldots, u(N-1)) \leq \varphi(\tilde{u}(0), \ldots, \tilde{u}(N-1))$ for all $\varphi(\tilde{u}(0), \ldots, \tilde{u}(N-1)) \in \Omega^N$.
If $\varphi(u(0), \ldots, u(N-1)) = 0$, then $\varphi(u(0), \ldots, u(N-1)) \in \Omega^N$ solves the equation (2.10). Otherwise no such solution exists. We again assume that $g \in C^1(\mathbb{R}^n \times \mathbb{R}^m)$. Let $\Omega \subseteq \mathbb{R}^n$ be open.

Then a necessary condition for $(u(0), \ldots, u(N-1)) \in \Omega^N$ to minimize φ on Ω^N is given by

$$\varphi_{u(k)}(u(0), \ldots, u(N-1)) = 2\, G^N_{u(k)}(x_0, u(0), \ldots, u(N-1))^T \times$$

$$(G^N(x_0, u(0), \ldots, u(N-1)) - \hat{x}) = \Theta_m \qquad (OC)$$

for all $k = 0, \ldots, N-1$.

For the determination of $(u(0), \ldots, u(N-1)) \in \Omega^N$ with (OC) one can apply *Marquardt's algorithm*: Let $(u(0), \ldots, u(N-1)) \in \Omega^N$ be chosen. If (OC) is satisfied, then $(u(0), \ldots, u(N-1))$ is taken as a solution of the optimization problem. Otherwise, for every $k \in \{0, \ldots, N-1\}$, a vector $h_\lambda(k) \in \mathbb{R}^m$ is determined *as solution of the linear system*

$$(2\, G^N_{u(k)}(x_0, u(0), \ldots, u(N-1))^T G^N_{u(k)}(x_0, u(0), \ldots, u(N-1)) + \lambda\, I_m h_\lambda(k)$$

$$= 2\, G^N_{u(k)}(x_0, u(0), \ldots, u(N-1))^T (G^N(x_0, u(0), \ldots, u(N-1)) - \hat{x})$$

where $\lambda > 0$ and I_m is the $m \times m$-unit matrix.

Then one can show (see, for instance [11]) that for sufficiently large $\lambda > 0$ it follows that

$$(u(0) + h_\lambda(0), \ldots, u(N-1) + h_\lambda(N-1)) \in \Omega$$

and

$$\varphi(u(0) + h_\lambda(0), \ldots, u(N-1) + h_\lambda(N-1)) < \varphi(u(0), \ldots, u(N-1)).$$

The algorithm is then continued with $(u(0) + h_\lambda(0), \ldots, u(N-1) + h_\lambda(N-1))$ instead of $(u(0), \ldots, u(N-1))$.

Now let us consider a special case which is motivated by a situation which occurs in the modelling of conflicts. We begin with an uncontrolled system of the form

$$x^1(t+1) = g_1(x^1(t), x^2(t))\,,$$
$$x^2(t+1) = g_2(x^1(t), x^2(t))\,, \quad t \in \mathbb{N}_0\,,$$

where $g_i : \mathbb{R}^{n_1} \times \mathbb{R}^{n_2} \to \mathbb{R}^{n_i}$, $i = 1, 2$, are given continuous mappings and $x^i : \mathbb{N}_0 \to \mathbb{R}^{n_i}$, $i = 1, 2$, are considered as state functions.

For $t = 0$ we assume initial conditions

$$x^1(0) = x_0^1\,, \quad x^2(0) = x_0^2 \qquad (2.11)$$

where $x_0^1 \in \mathbb{R}^{n_1}$ and $x_0^2 \in \mathbb{R}^{n_2}$ are given vectors with

$$\Theta_{n_2} \le x_0^2 \le x_*^2$$

for some $x_*^2 \geq \Theta_{n_2}$ which is also given. We further assume that the above system admits fixed points $(\hat{x}^{1^T}, \hat{x}^{2^T})^T \in \mathbb{R}^{n_1} \times \mathbb{R}^{n_2}$ with

$$\Theta_{n_2} \leq \hat{x}^2 \leq x_*^2$$

which are then solutions of the system

$$\hat{x}^1 = g_1(\hat{x}^1, \hat{x}^2) \ , \ \hat{x}^2 = g_2(\hat{x}^1, \hat{x}^2) \ .$$

Now we consider the following

Problem: Find vector functions $x^1 : \mathbb{N}_0 \to \mathbb{R}^{n_1}$ and $x^2 : \mathbb{N}_0 \to \mathbb{R}^{n_2}$ with

$$\Theta_{n_2} \leq x^2(t) \leq x_*^2 \ \text{ for all } \ t \in \mathbb{N}_0$$

which satisfy the above system equations and initial conditions and

$$x^1(t) = \hat{x}^1 \ , \ x^2(t) = \hat{x}^2 \ \text{ for all } \ t \geq N$$

where $N \in \mathbb{N}_0$ is a suitably chosen integer. In general this problem will not have a solution. Therefore we replace the uncontrolled system by the following controlled system:

$$\begin{aligned} x^1(t+1) &= g_1(x^1(t), x^2(t) + u(t)) \ , \\ x^2(t+1) &= g_2(x^1(t), x^2(t) + u(t)) \ , \ t \in \mathbb{N}_0 \ , \end{aligned} \qquad (2.12)$$

where $u : \mathbb{N}_0 \to \mathbb{R}^{n_2}$ is a control function. Then we consider the problem of finding a control function $u : \mathbb{N}_0 \to \mathbb{R}^{n_2}$ such that the solutions $x^1 : \mathbb{N}_0 \to \mathbb{R}^{n_1}$ and $x^2 : \mathbb{N}_0 \to \mathbb{R}^{n_2}$ of (2.12) and (2.11) satisfy the conditions

$$\Theta_{n_2} \leq x^2(t) + u(t) \leq x_*^2 \ \text{ for all } \ t \in \mathbb{N}_0$$

and

$$x^1(t) = \hat{x}^1 \ , \ x^2(t) + u(t) = \hat{x}^2 \ \text{ for all } \ t \geq N$$

where $N \in \mathbb{N}_0$ is a suitably chosen integer.

Let us assume that we can find a vector function $v : \mathbb{N}_0 \to \mathbb{R}^n$ with

$$\Theta_{n_2} \leq v(t) \leq x_*^2 \ \text{ for } \ t = 0, \ldots, N-1 \ \text{ and } \ v(t) = \hat{x}^2 \ \text{ for } \ t \geq N$$

such that the solution $x^1 : \mathbb{N}_0 \to \mathbb{R}^{n_1}$ of

$$\begin{aligned} x^1(t+1) &= g_1(x^1(t), v(t)) \ , \ t \in \mathbb{N}_0 \ , \\ x^1(0) &= x_0^1 \end{aligned}$$

satisfies

$$x^1(t) = \hat{x}^1 \ \text{ for all } \ t \geq N$$

where $N \in \mathbb{N}$ is a suitably chosen integer.

Then we put

$$x^2(0) = x_0^2 \,,$$
$$x^2(t+1) = g_2(x^1(t),\ v(t)) \quad \text{for all } t \in \mathbb{N}_0$$

and define

$$u(t) = v(t) - x^2(t) \quad \text{for } t \in \mathbb{N}_0.$$

With these definitions we obtain a solution of the above control problem. Thus in order to find such a solution we have to find a vector function $v : \mathbb{N}_0 \to \mathbb{R}^{n_2}$ with

$$\Theta_{n_2} \le v(t) \le x_*^2 \quad \text{for } t = 0, \dots, N-1 \,,$$
$$v(t) = \hat{x}^2 \quad \text{for } t \ge N$$

such that the solution $x^1 : \mathbb{N}_0 \to \mathbb{R}^{n_1}$ of

$$x^1(t+1) = g_1(x^1(t), v(t)) \,, \quad t \in \mathbb{N}_0 \,,$$
$$x^1(0) = x_0^1$$

satisfies

$$x^1(t) = \hat{x}^1 \quad \text{for all } t \ge N$$

where $N \in \mathbb{N}$ is a suitably chosen integer. Let us demonstrate all this by an emission reduction model (1.22) to which we add the conditions

$$0 \le M_i(t) \le M_i^* \quad \text{for all } t \in \mathbb{N}_0 \ \text{ and } \ i = 1, \dots, r$$

and the initial conditions

$$E_i(0) = E_{0i} \ \text{ and } \ M_i(0) = M_{0i} \quad \text{for } i = 1, \dots, r$$

where $E_{0i} \in \mathbb{R}$ and $M_{0i} \in \mathbb{R}$ with $0 \le M_{0i} \le M_i^*$ for $i = 1, \dots, r$ are given. The corresponding controlled system (2.12) reads in this case

$$E_i(t+1) = E_i(t) + \sum_{j=1}^{r} em_{ij}(M_j(t) + u_j(t)) \,,$$
$$M_i(t+1) = M_i(t) + u_i(t) - \lambda_i(M_i(t) + u_i(t))(M_i^* - M_i(t) - u_i(t))E_i(t)$$

for $i = 1, \dots, r$ and $t \in \mathbb{N}_0$.
The control functions $u_i : \mathbb{N}_0 \to \mathbb{R}$, $i = 1, \dots, r$, must satisfy the conditions

$$0 \le M_i(t) + u_i(t) \le M_i^* \quad \text{for } i = 1, \dots, r \ \text{ and } \ t \in \mathbb{N}_0.$$

Fixed points of the system (1.22) are of the form $(\hat{E}^T, \Theta_r^T)^T$ with $\hat{E} \in \mathbb{R}^r$ arbitrary. We have to find a vector function $v : \mathbb{N}_0 \to \mathbb{R}^r$ with

$$\Theta_r \le v(t) \le M^* \quad \text{for } t = 0, \dots, N-1 \,,$$
$$v(t) = \Theta_r \quad \text{for } t \ge N$$

such that the solution $E : \mathbb{N}_0 \to \mathbb{R}^r$ of

$$E(t+1) = E(t) + Cv(t) \ , \ t \in \mathbb{N}_0 \ , \ (C = (em_{ij})_{i,j=1,\ldots,r})$$
$$E(0) = E_0 \ ,$$

satisfies

$$E(t) = \hat{E} \ \text{ for all } \ t \geq N$$

where $N \in \mathbb{N}$ is a suitably chosen integer.
First of all we observe that for every $N \in \mathbb{N}$

$$E(N) = E_0 + C\left(\sum_{t=0}^{N-1} v(t)\right)$$

Let us assume that C is invertible and C^{-1} is positive. Further we assume that $\hat{E} \geq E_0$.
Then $E(N) = \hat{E}$, if and only if

$$\sum_{t=0}^{N-1} v(t) = C^{-1}(\hat{E} - E_0) \geq \Theta_r.$$

If we define

$$v(t) = \Theta_r \ \text{ for all } \ t \geq N \ ,$$

then

$$E(t) = \hat{E} \ \text{ for all } \ t \geq N.$$

Let us put

$$v_N = \sum_{t=0}^{N-1} v(t) = C^{-1}(\hat{E} - E_0).$$

If we define

$$v(t) = \frac{1}{N}v_N \ \text{ for } \ t = 0, \ldots, N-1$$

then

$$\sum_{t=0}^{N-1} v(t) = C^{-1}(\hat{E} - E_0)$$

and

$$\Theta_r \leq v(t) \leq M^* \ \text{ for } \ t = 0, \ldots, N-1$$

for sufficiently large N, if

$$M_i^* > 0 \ \text{ for } \ i = 1, \ldots, r.$$

We finish with a numerical example: $r = 3$, $E_0 = (0,0,0)^T$, $\hat{E} = (10, 10, 10)^T$, $M^* = (1, 1, 1)^T$, and

$$C = \begin{pmatrix} 1 & -0.8 & 0 \\ 0 & 1 & -0.8 \\ -0.1 & -0.5 & 1 \end{pmatrix}.$$

Then we have to solve the linear system

$$
\begin{aligned}
v_{N_1} - 0.8\, v_{N_2} &= 10 , \\
v_{N_2} - 0.8\, v_{N_3} &= 10 , \\
-0.1\, v_{N_1} - 0.5\, v_{N_2} + v_{N_3} &= 10 ,
\end{aligned}
$$

The solution reads

$$
\begin{aligned}
v_{N_1} &= 38.059701 , \\
v_{N_2} &= 35.074627 , \\
v_{N_3} &= 31.343284 .
\end{aligned}
$$

We choose $N = 39$. Then we have to put

$$v(t) = \frac{1}{N} v_N = \begin{pmatrix} 0.9758898 \\ 0.8993494 \\ 0.803674 \end{pmatrix} \quad \text{for } t = 0, \ldots, 38.$$

2.1.2 Null-Controllability of Linear Systems

Instead of (2.1) we consider a *linear system* of the form

$$x(t+1) = Ax(t) + Bu(t) , \quad t \in \mathbb{N}_0 , \tag{2.13}$$

where A is a real $n \times n$ − *matrix* and B a real $n \times m$ − *matrix* and where $u : \mathbb{N}_0 \to \mathbb{R}^m$ is a given control function. The corresponding *uncontrolled system* reads

$$x(t+1) = Ax(t) , \quad t \in \mathbb{N}_0 , \tag{2.14}$$

and admits $\hat{x} = \Theta_n$ as a fixed point.
The problem of fixed point controllability is then equivalent to the

Problem of Null-Controllability

Given $x_0 \in \mathbb{R}^n$ find some $N \in \mathbb{N}_0$ and a control function $u \in U$ (2.5) with (2.7) such that the solution $x : \mathbb{N}_0 \to \mathbb{R}^n$ of (2.13), (2.2) satisfies the end condition

$$x(N) = \Theta_n \tag{2.15}$$

(which implies $x(t) = \Theta_n$ for all $t \geq N$).

From (2.13) and (2.2) it follows that

$$x(N) = A^N x_0 + \sum_{t=1}^{N} A^{N-t} Bu(t-1) \tag{2.16}$$

so that (2.15) turns out to be equivalent to

$$\sum_{t=1}^{N} A^{N-t} Bu(t-1) = -A^N x_0 . \tag{2.17}$$

Now let A be non-singular. Then the set $S(\Theta_n)$ of all vectors $x_0 \in \mathbb{R}^n$ such that there exists a time $N \in \mathbb{N}$ and a solution $(u(0)^T, \ldots, u(N-1)^T)^T \in \Omega^N$ of the system (2.13) is given by

$$S(\Theta_n) = \bigcup_{N \in \mathbb{N}} E(N)$$

where, for every $N \in \mathbb{N}$,

$$E(N) = \{x = \sum_{t=1}^{N} A^{N-t} Bu(t-1) \mid u \in U \ (2.5)\}.$$

Next we assume that $\Omega \subseteq \mathbb{R}^m$ is convex, has Θ_m as interior point and satisfies

$$u \in \Omega \Rightarrow -u \in \Omega . \tag{2.18}$$

Then, for every $N \in \mathbb{N}$, the set $E(N)$ is convex and $E(N) = -E(N)$. This implies because of

$$E(N) \subseteq E(N+1) \quad \text{for all} \ \ N \in \mathbb{N}$$

that $S(\Theta_n)$ is also convex and $S(\Theta_n) = -S(\Theta_n)$.
Further we assume *Kalman's condition*, i.e. there exists some $N_0 \in \mathbb{N}$ such that

$$rank(B \mid AB \mid \ldots \mid A^{N_0-1} B) = n. \tag{2.19}$$

Then we can prove

Theorem 2.1. *If A is non-singular, $\Omega \subseteq \mathbb{R}^m$ is convex, has Θ_m as interior point and satisfies (2.18) and if Kalman condition (2.19) is satisfied for some $N_0 \in \mathbb{N}$, then Θ_n is an interior point of $S(\Theta_n)$.*

Proof. Let us assume that Θ_n is not an interior point of $S(\Theta_n)$. Then $S(\Theta_n)$ must be contained in a hyperplane through Θ_n, i.e. there must exist some $y \in \mathbb{R}^n$, $y \neq \Theta_n$, with

$$y^T x = 0 \quad \text{for all} \ \ x \in S(\Theta_n).$$

This implies

$$y^T(\sum_{t=1}^{N} A^{N-t} Bu(t-1)) = 0$$

for all $(u(0)^T, \dots, u(N-1)^T)^T \in \mathbb{R}^{m \cdot N}$ and all $N \in \mathbb{N}$,

hence

$$y^T A^{N-t} B = \Theta_m^T \text{ and all } N \in \mathbb{N}.$$

In particular for $N = N_0$ this implies $y = \Theta_n$ due to *Kalman condition* (2.19) which contradicts $y \neq \Theta_n$. Therefore the assumption that Θ_n is not an interior point of $S(\Theta_n)$ is false.

$\square$

In addition to the assumption of *Theorem 2.1* we assume that all the eigenvalues of A are less than 1 in absolute value. Then according to the *Corollary* following *Theorem 1.6* Θ_n is a global attractor of the *uncontrolled system* (2.14), i.e. $\lim_{t \to \infty} x(t) = \Theta_n$ where $x : \mathbb{N}_0 \to \mathbb{R}^n$ is a solution (2.14) with (2.2) for any $x_0 \in \mathbb{R}^n$. By *Proposition 2.1* therefore the problem of null-controllability has a solution for every choice of $x_0 \in \mathbb{R}^n$. If the set Ω of control vector values has the form

$$\Omega = \{u \in \mathbb{R}^m \mid \|u\| \leq \gamma\} \tag{2.20}$$

for some $\gamma > 0$ where $\|\cdot\|$ is any norm in $\mathbb{R}^m$, then this result can be strengthened to

Theorem 2.2. *Let the Kalman condition* (2.19) *be satisfied for some* $N_0 \in \mathbb{N}$. *Further let* Ω *be of the form* (2.20).
Finally let all the eigenvalues of A^T *be less than or equal to one in absolute value and the corresponding eigenvectors be linearly independent. Then the problem of null-controllability has a solution for every choice of* $x_0 \in \mathbb{R}^n$, *if* A *is non-singular.*

Proof. We have to show that for every choice of $x_0 \in \mathbb{R}^n$ there is some $N \in \mathbb{N}_0$ and a control function $u \in U$ (2.5) such that (2.17) is satisfied. Since A is non-singular, (2.17) is equivalent to

$$\sum_{t=1}^{N} A^{-t} Bu(t-1) = -x_0.$$

For every $N \in \mathbb{N}$ we define the convex set

$$R(N) = \{x = \sum_{t=1}^{N} A^{-t} Bu(t-1) \mid u \in U\}$$

and put

$$R_\infty = \bigcup_{N \in \mathbb{N}} R(N).$$

Because of

$$R(N) \subseteq R(N+1) \quad \text{for all} \ \ N \in \mathbb{N}_0$$

the set R_∞ is also convex. We have to show that $R_\infty = \mathbb{R}^n$. Let us assume that $R_\infty \neq \mathbb{R}^n$.

Then there exists some $\tilde{x} \in \mathbb{R}^n$ with $\tilde{x} \notin R_\infty$ which can be separated from R_∞ by a hyperplane, i.e., there exists a number $\alpha \in \mathbb{R}$ and a vector $y \in \mathbb{R}^n$, $y \neq \Theta_n$ such that

$$y^T x \leq \alpha \leq y^T \tilde{x} \quad \text{for all} \ \ x \in R_\infty.$$

Since $\Theta_n \in R_\infty$, it follows that $\alpha \geq 0$. Further it follows from the implication $u \in \Omega \ \Rightarrow \ -u \in \Omega$ that

$$\left| \sum_{t=1}^{N} y^T A^{-t} B u(t-1) \right| \leq \alpha \quad \text{for all} \ \ N \in \mathbb{N} \ \ \text{and all} \ \ u \in U .$$

This implies

$$\sum_{t=1}^{N} \|(y^T A^{-t} B)^T\|_d \leq \alpha \quad \text{for all} \ \ N \in \mathbb{N}$$

where $\| \cdot \|_d$ is the norm in $\mathbb{R}^m$ which is dual to $\| \cdot \|$. This in turn implies

$$\lim_{t \to \infty} y^T A^{-t} B = \Theta_m^T. \tag{2.21}$$

From *Kalman's condition* (2.19) it follows that there exist n linearly independent vectors in $\mathbb{R}^n$ of the form

$$c_i = A^{t_i} b_{j_i} \quad \text{for} \ \ i = 1, \ldots, n$$

where $t_i \in \{0, \ldots, N_0 - 1\}$ and $j_i \in \{1, \ldots, m\}$ and b_{j_i} denotes the $j_i - th$ column vector of B. From (2.21) it follows that

$$\lim_{t \to \infty} y^T A^{-t} c_i = 0 \quad \text{for} \ \ i = 1, \ldots, n.$$

This implies

$$\lim_{t \to \infty} y^T A^{-t} = \Theta_n^T$$

or, equivalently,

$$\lim_{t \to \infty} (A^T)^{-t} y = \Theta_n. \tag{2.22}$$

Now let $\lambda_1, \ldots, \lambda_n \in \mathbb{C}$ be the eigenvalues of A^T and $y_1, \ldots, y_n \in \mathbb{C}^n$ corresponding linearly independent eigenvectors. Then there is a unique representation

$$y = \sum_{j=1}^{n} \alpha_j \, y_j \quad \text{where not all } \alpha_j \in \mathbb{C} \text{ are zero}$$

and

$$\left(A^T\right)^{-t} y_j = \left(\frac{1}{\lambda_j}\right)^t y_j \quad \text{for } j = 1, \ldots, n \,,$$

hence

$$\left(A^T\right)^{-t} y = \sum_{j=1}^{n} \alpha_j \left(\frac{1}{\lambda_j}\right)^t y_j \quad \text{for all } t \in \mathbb{N},$$

From (2.22) we therefore infer that

$$|\lambda_j| > 1 \quad \text{for all } j \in \{1, \ldots, n\} \text{ with } \alpha_j \neq 0.$$

This is a contradiction to

$$|\lambda_j| \leq 1 \quad \text{for all } 1, \ldots, n.$$

Hence the assumption $R_\infty \neq R^n$ is false.

$\square$

Remark: If we define

$$\overline{Y} = (y_1 \mid y_2 \mid \ldots \mid y_n) \quad \text{and } \wedge = \begin{pmatrix} \lambda_1 & & 0 \\ & \ddots & \\ 0 & & \lambda_n \end{pmatrix},$$

then it follows that

$$A^T \overline{Y} = \overline{Y} \wedge$$

which implies

$$\overline{Y}^{-1} A^T \overline{Y} = \wedge$$

and in turn

$$\overline{Y}^T A (\overline{Y}^T)^{-1} = \wedge \quad (\text{ since } (A^{-1})^T = (A^T)^{-1})$$

from which

$$A (\overline{Y}^T)^{-1} = (\overline{Y}^T)^{-1} \wedge$$

follows.

This implies that A has the same eigenvalues as A^T (which holds for arbitrary matrices) and the eigenvectors of A are the column vectors of $(\overline{Y}^T)^{-1}$. Therefore A^T in *Theorem 2.2* could be replaced by A. For the following let us assume that $\Omega = \mathbb{R}^m$.

For every $N \in \mathbb{N}$ let us define

$$\overline{Y}(N) = (B \mid AB \mid \ldots \mid A^{N-1}B).$$

Since U (2.5) consists of all functions $u : \mathbb{N}_0 \to \mathbb{R}^m$, it follows, for every $N \in \mathbb{N}$, that

$$E(N) = \{x = \sum_{t=1}^{N} A^{N-t} Bu(t-1) \mid u : \mathbb{N}_0 \to \mathbb{R}^m\}.$$

Further we can prove

Proposition 2.2. *The following statements are equivalent:*

(i) $rank\ \overline{Y}(N) = rank\ \overline{Y}(N+1)$;

(ii) $E(N) = E(N+1)$;

(iii) $(A^N B)\mathbb{R}^m \subseteq E(N)$;

(iv) $rank\ \overline{Y}(N) = rank\ \overline{Y}(N+j)$ for all $j \geq 1$.

Proof.

(i) $\Rightarrow$ (ii) : This is a consequence of the fact that $E(N) \subseteq E(N+1)$.

(ii) $\Rightarrow$ (iii) : This follows from $\overline{Y}(N+1) = (\overline{Y}(N) \mid A^N B)$.

(iii) $\Rightarrow$ (i) : $\overline{Y}(N+1) = (\overline{Y}(N) \mid A^N B)$ shows that (iii) $\Rightarrow$ (ii) and obviously we have (ii) $\Rightarrow$ (i).

(i) $\Rightarrow$ (iv) : Since (i) implies (iii), it follows that

$$(A^{N+1}B)\mathbb{R}^m \subseteq AE(N) \subseteq E(N+1)$$

which implies $E(N+1) = E(N+2)$ and hence

$$rank(\overline{Y}(N+1) = rank(\overline{Y}(N+2).$$

(iv) $\Rightarrow$ (i) : is obvious. This completes the proof.

$\square$

Now let r be the smallest integer such that $I, A, \ldots, A^{r-1}$ are linearly independent in $\mathbb{R}^{n \cdot n}$ and hence there are numbers $\alpha_{r-1}, \alpha_{r-2}, \ldots, \alpha_0 \in \mathbb{R}$ such that

$$A^r + \alpha_{r-1} A^{r-1} + \ldots + \alpha_0 I = 0.$$

Defining

$$\Phi_0(\lambda) = \lambda^r + \alpha_{r-1} \lambda^{r-1} + \ldots + \alpha_0 ,$$

we have $\Phi_0(A) = 0$. This monic polynomial (leading coefficient 1) is the monic polynomial of least degree for which $\Phi_0(A) = 0$ and is called the *minimal polynomial* of A. The polynomial

$$\Phi(\lambda) = det(\lambda I - A) \quad \text{with degree } n$$

is called *characteristic polynomial* of A, and the *Hamilton-Cayley Theorem* states that $\Phi(A) = 0$ which implies $r \leq n$.
This leads to

Proposition 2.3. *Let s be the degree of the minimal polynomial of $A(s \leq n)$. Then there is an integer $k \leq s$ such that*

$$rank\overline{Y}(1) < rank\overline{Y}(2) < \ldots < rank\overline{Y}(k) = rank\overline{Y}(k+j) \quad \text{for all } j \in \mathbb{N}$$

Proof. *Proposition 2.2* implies the existence of such an integer k, since $rank\overline{Y}(N) \leq n$ for all $N \in \mathbb{N}$. We have to show that $k \leq s$. Let $\psi(\lambda) = \lambda^s + \alpha_{s-1} \lambda^{s-1} + \ldots + \alpha_0$ be the minimal polynomial of A. Then $\psi(A)B = 0$ and $A^s B \mathbb{R}^m \subseteq E(s)$ which implies (by *Proposition 2.2*) that $rank\overline{Y}(s) = rank\overline{Y}(s+j)$ for all $j \in \mathbb{N}$, hence $k \leq s$.

$\square$

As a consequence of *Proposition 2.3* we obtain

Proposition 2.4. *If Kalman's condition (2.19) is satisfied for some $N_0 \in \mathbb{N}$, then necessarily $N_0 \leq n$ and*

$$rank(B \mid AB \mid \ldots \mid A^{n-1} B) = n ,$$

hence $E(n) = \mathbb{R}^n$. Conversely, if $E(n) = \mathbb{R}^n$, then Kalman's condition (2.19) is satisfied for all $N \geq n$ and

$$E(N) = \mathbb{R}^n \quad \text{for all } N \geq n.$$

Proposition 2.3 also implies that, if $rank\overline{Y}(n) < n$, then $rank\overline{Y}(N) < n$ for all $N \geq n$ and

$$E(N) = E(n) \neq \mathbb{R}^n \quad \text{for all } N \geq n.$$

If we define, for ever $N \in \mathbb{N}$, the $n \times n$-matrix

$$W(N) = \overline{Y}(N)\overline{Y}(N)^T = \sum_{j=0}^{N-1} A^j BB^T (A^j)^T ,$$

then it follows that

$$W(N)\mathbb{R}^n \subseteq E(N) \quad \text{for every } N \in \mathbb{N}.$$

Now let us assume that $rank \; \overline{Y}(N) = n$ which is equivalent to $E(N) = \mathbb{R}^n$. Then $W(N)$ is non-singular and

$$W(N)\mathbb{R}^n = \mathbb{R}^n = E(N).$$

Let $y^* \in \mathbb{R}^n$ be the unique solution of

$$W(N)y^* = -A^N x_0.$$

Further let $u \in \mathbb{R}^{m \cdot N}$ be any solution of

$$\overline{Y}(N)u = -A^N x_0 \qquad \text{(see (2.17))}.$$

If we put

$$u^* = \overline{Y}(N)^T y^* \quad (\in \mathbb{R}^{m \cdot N}) ,$$

then

$$\overline{Y}(N)u^* = -A^N x_0$$

and

$$\|u^*\|_2^2 = {u^*}^T u^* = {y^*}^T \overline{Y}(N)u^* = {y^*}^T W(N)y^*$$

$$= -{y^*}^T A^N x_0 = y^* \overline{Y}(N)u = {u^*}^T u \leq \|u^*\|_2 \|u\|_2$$

which implies $\|u^*\|_2 \leq \|u\|_2$.

2.1.3 A Method for Solving the Problem of Null-Controllability

Let us equip $\mathbb{R}^m$ with the *Euclidean norm* $\|\cdot\|_2$ and consider the following

Problem (P)

For a given $N \in \mathbb{N}$ find $u : \{0,\ldots,N-1\} \to \mathbb{R}^m$ such that

$$\sum_{t=1}^{N} A^{N-t} B u(t-1) = -A^N x_0 \tag{2.23}$$

and

$$\varphi_N(u) = \max_{t=1,\ldots,N} \|u(t-1)\|_2$$

is as small as possible.

If Kalman's condition (2.19) is satisfied for some $N_0 \in \mathbb{N}$ and if $N \geq N_0$, then *Problem (P)* has a solution $u_N \in \mathbb{R}^{m \cdot N}$. If $\varphi_N(u_N) \leq \gamma$ (see (2.20)), then we obtain solution $u_N : \mathbb{N}_0 \to \mathbb{R}^m$ of the problem of *null-controllability* if we define

$$u_N(t) = \Theta_m \quad \text{for all } t \geq N.$$

If $\varphi_N(u_N) > \gamma$, then N must be increased.

If the matrix A is non-singular, then (2.17) is equivalent to

$$\sum_{t=1}^{N} A^{-t} B u(t-1) = -x_0$$

which implies

$$\varphi_{N+1}(u_{N+1}) \leq \varphi_N(u_N) \quad \text{for all } N \geq N_0.$$

Under the assumptions of *Theorem 2.2* there exists, for every $\varepsilon > 0$, some $N(\varepsilon) \in \mathbb{N}$ such that

$$\varphi_{N(\varepsilon)}(u_{N(\varepsilon)}) \leq \varepsilon$$

which implies

$$\lim_{N \to \infty} \varphi_N(u_N) = 0.$$

So we can be sure, for every choice of $\gamma > 0$, to find a solution $u_{N(\gamma)} \in \mathbb{R}^{m \cdot N(\gamma)}$ of *Problem (P)* with $\varphi_{N(\gamma)}(u_{N(\gamma)}) \leq \gamma$ which leads to a solution $u_{N(\gamma)} : \mathbb{N}_0 \to \mathbb{R}^m$ of the problem of *null-controllability*, if we define

$$u_{N(\gamma)}(t) = \Theta_m \quad \text{for all } t \geq N(\gamma).$$

In order to solve *Problem (P)* we replace it by

Problem (D)

Minimize

$$\chi(y) = \sum_{k=1}^{N} \|B^T (A^{N-k})^T y\|_2 , \ y \in \mathbb{R}^n ,$$

subject to

$$c^T y = -x_0^T (A^T)^N y = 1. \qquad (2.24)$$

Let $u : \{0,\ldots,N-1\} \to \mathbb{R}^m$ be a solution of (2.17) and let $y \in \mathbb{R}^n$ satisfy (2.24). Then it follows that

$$\sum_{k=1}^{N} y^T A^{N-k} B u(k-1) = y^T c = 1$$

which implies

$$\max_{k=1,\ldots,N} \|u(k-1)\|_2 \geq \frac{1}{\chi(y)}.$$

Now let $\hat{y} \in \mathbb{R}^n$ be a solution of *Problem (D)* . Then there is a multiplier $\lambda \in \mathbb{R}$ such that

$$\nabla \chi(\hat{y}) = \lambda c \qquad (2.25)$$

$$\nabla \chi(\hat{y}) = \sum_{k \in I(\hat{y})} \frac{1}{\|B^T (A^{N-k})^T \hat{y}\|_2} A^{N-k} B B^T (A^{N-k})^T \hat{y}$$

with

$$I(\hat{y}) = \{k \mid \|B^T (A^{N-k})^T \hat{y}\|_2 > 0\}.$$

This implies

$$\lambda = \hat{y}^T \nabla \chi(\hat{y}) = x(\hat{y}).$$

If we define

$$u_N(k-1) = \begin{cases} \frac{1}{x(\hat{y})} \frac{1}{\|B^T (A^{N-k})^T \hat{y}\|_2} B^T (A^{N-k})^T \hat{y} , & \text{if } k \in I(\hat{y}), \\ \Theta_m & \text{else} , \ k = 1,\ldots,N , \end{cases}$$

$$(2.26)$$

then it follows that

$$\sum_{k=1}^{N} A^{N-k} B u_N(k-1) = c$$

and

$$\|u_N(k-1)\|_2 = \frac{1}{\chi(\hat{y})} \quad \text{for all } k \in I(\hat{y})$$

which implies

$$\max_{k=1,\ldots,N} \|u_N(k-1)\|_2 = \frac{1}{\chi(\hat{y})}.$$

Hence $u_N : \{0,\ldots,N-1\} \to \mathbb{R}^m$ solves *Problem (P)*. This result is summarized as

Theorem 2.3. *If $\hat{y} \in \mathbb{R}^n$ solves Problem (D), then $u_N : \{0, \ldots, N-1\} \to \mathbb{R}^m$ defined by (2.26) solves Problem (P).*

In order to solve *Problem (D)* we apply the well known *gradient projection method* which is based on the following iteration step: Let $y^* \in \mathbb{R}^n$ with $c^T y^* = 1$ be givenn. (At the beginning we take $y^* = \frac{1}{\|c\|_2^2} c$). Then we calculate

$$h = (\frac{1}{\|c\|_2^2} c^T \nabla \chi(y^*)) c - \nabla \chi(y^*)$$

and see that

$$ch = 0$$

and

$$\nabla \chi(y^*)^T h = \frac{1}{\|c\|_2^2} (c^T \nabla \chi(y^*))^2 - \|\nabla \chi(y^*)\|_2^2 \le 0 .$$

If $\nabla \chi(y^*)^T h = 0$, then there exists some $\lambda \ge 0$ such that (2.25) holds true which is equivalent to y^* being optimal. If $\nabla x(y^*)^T h < 0$, then h is a *feasible direction of descent*. If we determine $\hat{\lambda} > 0$ such that

$$\chi(y^* + \hat{\lambda} h) = \min_{\lambda > 0} \chi(y^* + \lambda h) , \qquad (2.27)$$

then

$$\chi(y^* + \hat{\lambda} h) < \chi(y^*)$$

and $c^T (y^* + \hat{\lambda} h) = 1$. The next step is then performed with $y^* + \hat{\lambda} h$ instead of y^*. A necessary and sufficient condition for $\hat{\lambda} > 0$ to satisfy (2.27) is

$$\frac{d}{d\lambda} \chi(y^* + \hat{\lambda} h) = \nabla \chi(y^* + \hat{\lambda} h)^T h = 0$$

which is equivalent to

$$\sum_{k \in I(y^* + \hat{\lambda} h)} \frac{1}{\|B^T (A^{N-k})^T (y^* + \hat{\lambda} h)\|_2} h^T A^{N-k} B B^T (A^{N-k})^T (y^* + \hat{\lambda} h) = 0$$

and in turn to the fixed point equation

$$\hat{\lambda} = \psi(\hat{\lambda})$$

with

$$\psi(\lambda) = \frac{\displaystyle\sum_{k \in I(y^* + \lambda h)} \frac{1}{\|B^T (A^{N-k})^T (y^* + \lambda h)\|_2} h^T A^{N-k} B B^T (A^{N-k})^T y^*}{\displaystyle\sum_{k \in I(y^* + \lambda h)} \frac{1}{\|B^T (A^{N-k})^T (y^* + \lambda h)\|_2} h^T A^{N-k} B B^T (A^{N-k})^T h}$$

In order to solve this equation we apply the iteration procedure

$$\lambda_{k+1} = \psi(\lambda_k), \ k \in \mathbb{N}_0$$

starting with $\lambda_0 = 0$.

Let us return to the problem of *fixed point controllability* in *Section 2.1.1.* and let us assume that $g : \mathbb{R}^n \times \mathbb{R}^m \to \mathbb{R}^n$ is continuously Fréchet differentiable. Then it follows that

$$G^N(x_0, u(0), \ldots, u(N-1)) - \hat{x} =$$

$$G^N(x_0, u(0), \ldots, u(N-1)) - G^N(\hat{x}, \Theta_m, \ldots, \Theta_m)$$

$$\approx J_{G^N}^x(\hat{x}, \Theta_m, \ldots, \Theta_m)(x_0 - \hat{x}) + \sum_{k=1}^{N} J_{G^N}^{u(k-1)}(\hat{x}, \Theta_m, \ldots, \Theta_m)u(k-1)$$

$$= J_g^x(\hat{x}, \Theta_m)^N(x_0 - \hat{x}) + \sum_{k=1}^{N} J_g^x(\hat{x}, \Theta_m)^{N-k} J_g(\hat{x}, \Theta_m)u(k-1)$$

where

$$J_g^x(\hat{x}, \Theta_m) = (g_{ix_j}(\hat{x}, \Theta_m))_{i,j=1,\ldots,n}$$

and

$$J_g^u(\hat{x}, \Theta_m) = (g_{iu_k}(\hat{x}, \Theta_m))_{\substack{i=1,\ldots,n \\ k=1,\ldots,m}}.$$

Therefore we replace equation (2.10) by

$$\sum_{k=1}^{N} J_g^x(\hat{x}, \Theta_m)^{N-k} J_g(\hat{x}, \Theta_m)u(k-1) = -J_g^x(\hat{x}, \Theta_m)^N(x_0 - \hat{x}) \qquad (2.28)$$

and solve the problem of finding $u : \{0, \ldots, N-1\} \to \mathbb{R}^m$ which solves (2.40) and minimizes

$$\varphi_N(u) = \max_{k=1,\ldots,N} \|u(k-1)\|_2.$$

Such a $u : \{0, \ldots, N-1\} \to \mathbb{R}^m$ is then taken as an approximate solution of (2.10).

The above problem has the form of *Problem (P)* at the beginning of this Section and can be solved by the method described above.

Finally we consider a special case in which the problem of fixed point controllability is reduced to a sequence of such problems which can be solved more easily.

For this purpose we consider the system

$$x(t+1) = g_0(x(t)) + \sum_{j=1}^{m} g_j(x(t))u_j(t) \ , \ t \in \mathbb{N}_0 \ , \tag{2.29}$$

where $g_j : \mathbb{R}^n \to \mathbb{R}^n$, $j = 1, \ldots, m$, are continuous vector functions.
For every control function $u : \mathbb{N}_0 \to \mathbb{R}^m$ there is exactly one function $x : \mathbb{N}_0 \to \mathbb{R}^m$ which satisfies (2.29) and the initial condition

$$x(0) = x_0 \ , \ x_0 \in \mathbb{R}^n \ \text{given.} \tag{2.30}$$

We denote it by $x = x(u)$. We assume that the uncontrolled system

$$x(t+1) = g_0(x(t)) \ , \ t \in \mathbb{N} \ ,$$

has a fixed point $\hat{x} \in \mathbb{R}^n$ which then solves the system

$$\hat{x} = g_0(\hat{x}) \ .$$

We again assume that the set U of admissible control functions is given by
(2.5) where $\Omega \subseteq \mathbb{R}^m$ is a subset with $\Theta_m \in \Omega$.
Let us define

$$\tilde{g}_0(x) = g_0(x) - x \ , \ x \in \mathbb{R}^n \ .$$

Then (2.29) can be rewritten in the form

$$x(t+1) = x(t) + \tilde{g}_0(x(t)) + \sum_{j=1}^{m} g_j(x(t))u_j(t) \ , \ t \in \mathbb{N}_0 \ . \tag{2.31}$$

In order to find some $N \in \mathbb{N}_0$ and a control function $u \in U$ with (2.7) such
that the solution $x : \mathbb{N}_0 \to \mathbb{R}^n$ of (2.30), satisfies the end condition (2.8) we
apply an iterative method. Starting with some $N_0 \in \mathbb{N}_0$ and some $u^0 \in U$ (for
instance $u^0(t) = \Theta_m$ for all $t \in \mathbb{N}_0$) we construct a sequence $(N_k)_{k \in \mathbb{N}}$ in $\mathbb{N}_0$
and a sequence $(u^k)_{k \in \mathbb{N}}$ in U as follows:
If $N_{k-1} \in \mathbb{N}_0$ and $u^{k-1} \in U$ are determined, we calculate $x(u^{k-1}) : \mathbb{N}_0 \to \mathbb{R}^n$
as the solution of (2.2) and (2.31) for $u = u^{k-1}$. Then we determine $N_k \in \mathbb{N}_0$
and $u^k \in U$ such that

$$u^k(t) = \Theta_m \text{ for all } t \geq N_k \tag{2.7$_k$}$$

and the solution $x(u^k) : \mathbb{N}_0 \to \mathbb{R}^n$ of (2.2) and

$$x(u^k)(t+1) = x(u^k)(t) + \tilde{g}_0(x(u^{k-1})(t))$$

$$+ \sum_{j=1}^{m} g_j(x(u^{k-1})(t))u_j^{k+1}(t) \ , \ t \in \mathbb{N}_0 \ ,$$

$$\tag{2.31$_k$}$$

satisfies the end condition

$$x(u^k)(N_k) = \hat{x} \ . \tag{2.8$_k$}$$

If we put

$$x^k = x_0 + \sum_{t=1}^{N_k} \tilde{g}_0(x(u^{k-1})(t-1))$$

and

$$B^k(t-1) = (g_1(x(u^{k-1})(t-1) \mid \ \ldots \ \mid g_m(x(u^{k-1})(t-1)) \ ,$$

then the end condition (2.8)$_k$ is equivalent to

$$\sum_{t=1}^{N} B^k(t-1)u^{k-1}(t-1) = \hat{x} - x^k \ .$$

2.1.4 Stabilization of Controlled Systems

Let $g : \mathbb{R}^n \times \mathbb{R}^m \to \mathbb{R}^n$ be a continuous mapping and let H be a family of continuous mappings $h : \mathbb{R}^n \to \mathbb{R}^m$. If we define, for every $h \in H$, the mapping $f_h : \mathbb{R}^n \to \mathbb{R}^n$ by

$$f_h(x) = g(x, h(x)) \ , \quad x \in \mathbb{R}^n \ ,$$

then f_h is continuous and $(\mathbb{R}^n, f_h)$ is a *time - discrete autonomous dynamical system*. Let $\hat{x} \in \mathbb{R}^n$ be a fixed point of

$$f(x) = g(x, \Theta_m) \ , \quad x \in \mathbb{R}^n.$$

Further we assume that

$$h(\hat{x}) = \Theta_m \ \text{ for all } \ h \in H$$

which implies that $\hat{x}$ is a fixed point of all f_h, $h \in H$. After these preparations we can formulate the

Problem of Stabilization

Find $h \in H$ such that $\{\hat{x}\}$ is asymptotically stable with respect to f_h.
We assume that $g : \mathbb{R}^n \times \mathbb{R}^m \to \mathbb{R}^n$ and every mapping $h \in H$ are continuously *Fréchet* differentiable. Then every mapping $f_h : \mathbb{R}^n \to \mathbb{R}^n$, $h \in H$, is also continuously *Fréchet* differentiable and, for every $x \in \mathbb{R}^n$, its Jacobi matrix is given by

$$J_{f_h}(x) = J_g^x(x, h(x)) + J_g^u(x, h(x))J_h^x(x) \qquad \text{where}$$
$$J_g^x(x, h(x)) = (g_{ix_j}(x, h(x)))_{\substack{i = 1, \ldots, n \\ j = 1, \ldots, n}} \qquad \text{and}$$
$$J_g^u(x, h(x)) = (g_{iu_k}(x, h(x)))_{\substack{i = 1, \ldots, n \\ k = 1, \ldots, m}} \ , J_h^x(x) = (h_{ix_j}(x))_{\substack{i = 1, \ldots, m \\ j = 1, \ldots, n}}$$

From the *Corollary of Theorem 1.5* we then obtain the

Theorem 2.4.

(a) Let the spectral radius $\varrho(J_{f_h}(\hat{x})) < 1$. Then $\hat{x}$ is asymptotically stable with respect to f_h.

(b) Let $(J_{f_h}(\hat{x}))$ be invertible and let all the eigenvalues of $\varrho(J_{f_h}(\hat{x}))$ be larger than 1 in absolute value. Then $\hat{x}$ is unstable with respect to f_h.

Special cases:

(a) Let

$$g(x, u) = Ax + Bu \; , x \in \mathbb{R}^n \; , \; u \in \mathbb{R}^m \; ,$$

where A is a real $n \times n$-matrix and B a real $n \times m$-matrix, respectively. Further let H be the family of all linear mapping $h : \mathbb{R}^n \to \mathbb{R}^m$ which are given by

$$h(x) = Cx \; , \; x \in \mathbb{R}^n \; ,$$

where C is an arbitrary real $m \times n$-matrix, respectively. If we choose $\hat{x} = \Theta_n$, then

$$f(\Theta_n) = g(\Theta_n, \Theta_m) = \Theta_n$$

and

$$h(\Theta_n) = \Theta_m \quad \text{for all} \; \; h \in H.$$

Finally we have $J_h(x) = C$,

$$J_g^x(x, h(x)) = A \; , \quad \text{and} \; \; J_g^u(x, h(x)) = B$$

for all $x \in \mathbb{R}^n$ and $h \in H$ which implies

$$J_{f_h}(x) = A + B \, C \quad \text{for all} \; \; x \in \mathbb{R}^n \; \; \text{and} \; \; h \in H.$$

Thus $\hat{x} = \Theta_n$ is asymptotically stable with respect to f_h, if

$$\varrho(A + B \, C) < 1,$$

and unstable with respect to f_h, if all the eigenvalues of $A + B \, C$ are larger than one in absolute value.

(b) Let

$$g(x, u) = F(x) + B(x)u \ , \ x \in \overline{X} \ , \ u \in \mathbb{R}^m \ ,$$

where $F : \overline{X} \to \overline{X}$, $\overline{X} \subseteq \mathbb{R}^n$ open, is continuously Fréchet differentiable and $B(x) = (b_1(x), \dots, b_m(x))$, $x \in \overline{X}$, where $b_j : \overline{X} \to \mathbb{R}^n, j = 1, \dots, n$, are also continuously Fréchet differentiable. Let again H be the family of all linear mappings $h : \mathbb{R}^n \to \mathbb{R}^m$ which are given by

$$h(x) = Cx \ , \ x \in \mathbb{R}^n.$$

Finally, we assume that $\Theta_n \in \overline{X}$ and $F(\Theta_n) = \Theta_n$.
If we choose $\hat{x} = \Theta_n$, then

$$f(\Theta_n) = g(\Theta_n, \Theta_m) = \Theta_n$$

and

$$h(\Theta_n) = \Theta_m \ \text{ for all } \ h \in H.$$

Further we obtain

$$J_h^x(x) = C \ ,$$

$$J_g^x(x, h(x)) = J_F(x) + \sum_{j=1}^m J_{b_j}(x)h_j(x) \ \text{ and } \ J_g^u(x, h(x)) = B(x) \ ,$$

for all $x \in \overline{X}$ and $h \in H$ which implies

$$J_{f_h}(x) = J_F(x) + \sum_{j=1}^m J_{b_j}(x)h_j(x) + B(x)C \ \text{ for all } \ x \in \overline{X} \ \text{ and } \ h \in H \ ,$$

hence

$$J_{f_h}(\Theta_n) = J_F(\Theta_n) + B(\Theta_n)C \ \text{ for all } \ h \in H.$$

Thus $\hat{x} = \Theta_n$ is asymptotically stable with respect to f_h, if

$$\varrho(J_F(\Theta_n) + B(\Theta_n)C) < 1$$

and unstable with respect to f_h, if all the eigenvalues of $J_F(\Theta_n) + B(\Theta_n)C$ are larger than one in absolute value.

2.1.5 Applications

a) An Emission Reduction Model:

We pick up the emission reduction model that was treated as uncontrolled system in *Section 1.1.6* and as controlled system in *Section 2.1.1*. Here we concentrate on the controlled system which we linearize at a fixed point $(\hat{E}^T, \theta_r^T)^T$, $\hat{E} \in \mathbb{R}^r$, of the uncontrolled system which leads to a linear control system of the form

$$x(t+1) = Ax(t) + Bu(t) , \quad t \in \mathbb{N}_0 ,$$

with

$$A = \begin{pmatrix} I_r & C \\ 0_r & D \end{pmatrix} , \quad B = \begin{pmatrix} C \\ D \end{pmatrix} ,$$

where I_r and 0_r is the $r \times r$-unit and zero-matrix, respectively, and

$$C = \begin{pmatrix} em_{11} & \dots & em_{1r} \\ \vdots & & \vdots \\ em_{r1} & \dots & em_{rr} \end{pmatrix} ,$$

$$D = \begin{pmatrix} 1 - \lambda_1 M_1^* \hat{E}_1 & & 0 \\ & \ddots & \\ 0 & & 1 - \lambda_r M_r^* \hat{E}_r \end{pmatrix} .$$

This implies

$$A^k B = \begin{pmatrix} C(I_r + D + \dots + D^k) \\ D^{k+1} \end{pmatrix} \quad \text{for all } k \in \mathbb{N}_0 .$$

We consider the problem of null-controllability as being discussed in *Section 2.1.2*. Let us assume that C and D are non-singular. Then it follows that the matrices A and

$$\begin{pmatrix} C & C(I_r + D) \\ D & D^2 \end{pmatrix}$$

are non-singular which implies that the Kalman condition (2.19) is satisfied for $N_0 = 2$. Let $d_1, \dots, d_r$ be the diagonal elements of D. Thus the non-singularity of D is equivalent to

$$d_i \neq 0 \quad \text{for all } i = 1, \dots, r.$$

If all $d_i \neq 1$ for $i = 1, \dots, r$, then it follows (see *Section 1.1.6*) that the eigenvectors corresponding to the eigenvalues

$$\mu_i = 1 \quad \text{for } i = 1, \dots, r \quad \text{and} \quad \mu_{i+r} = d_i \quad \text{for } i = 1, \dots, r$$

of A are linearly independent which also holds true for A^T (which has the same eigenvalues)(see *Section 2.1.2*).

If

$$|d_i| \leq 1 \quad \text{for all} \quad i = 1, \ldots, r$$

and

$$\Omega = \{u \in \mathbb{R}^r \mid \|u\| \leq \gamma\}$$

for some $\gamma > 0$ where $\|\cdot\|$ is any norm in $\mathbb{R}^r$, then by *Theorem 2.2* the problem of null-controllability has a solution for every choice of $x_0 = (x_0^{1^T}, x_0^{2^T}) \in \mathbb{R}^{2r}$. This problem can be solved with the aid of *Problem (P)* in *Section 2.1.3* which reads as follows in this case: For a given $N \in \mathbb{N}$ find $u : \{0, \ldots, N-1\} \to \mathbb{R}^r$ such that

$$\sum_{k=1}^{N} \left(\frac{C(I_r + D + \ldots + D^k)}{D^{k+1}} \right) u(k-1) =$$

$$- \left(\begin{matrix} I_r & C(I_r + D + \ldots + D^{N-1}) \\ 0_r & D^N \end{matrix} \right) \begin{pmatrix} x_0^1 \\ x_0^2 \end{pmatrix}$$

and

$$\varphi_N(u) = \max_{k=1,\ldots,N} \|u(k-1)\|_2$$

is minimized (where $\|\cdot\|_2$ denotes the *Euclidean* norm in $\mathbb{R}^r$).

Finally we illustrate the method by two numerical examples.

Let $m = 3$. In both cases with choose

$$x_0^{1^T} = x_0^{2^T} = (1, 1, 1) \tag{2.32}$$

At first we choose

$$C = \begin{pmatrix} 0.8 & 0.5 & -0.5 \\ 0.2 & 0.2 & 0.3 \\ 0.4 & 0.3 & 0.2 \end{pmatrix}$$

and

$$D = \begin{pmatrix} 0.1 & 0 & 0 \\ 0 & -0.2 & 0 \\ 0 & 0 & 0.1 \end{pmatrix}$$

and obtain

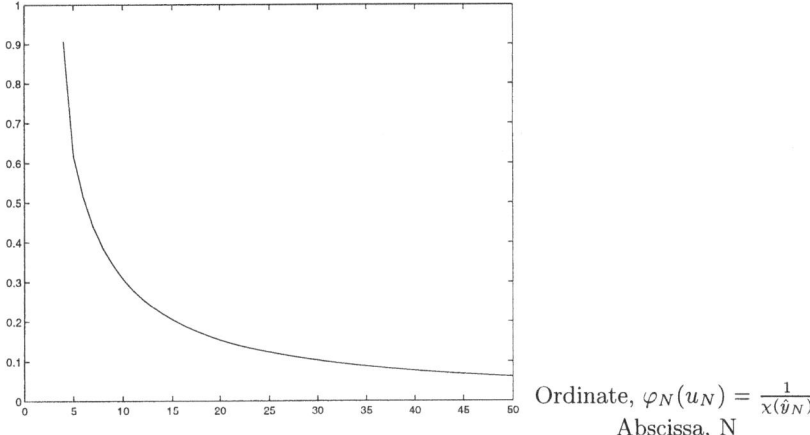

Ordinate, $\varphi_N(u_N) = \frac{1}{\chi(\hat{y}_N)}$
Abscissa, N

Next we choose

$$C = \begin{pmatrix} 0.8 & 0.5 & 0.5 \\ 0.2 & 0.2 & 0.3 \\ 0.4 & 0.1 & 0.2 \end{pmatrix} \qquad D = \begin{pmatrix} 0.2 & 0 & 0 \\ 0 & 0.8 & 0 \\ 0 & 0 & 0.6 \end{pmatrix}$$

and get

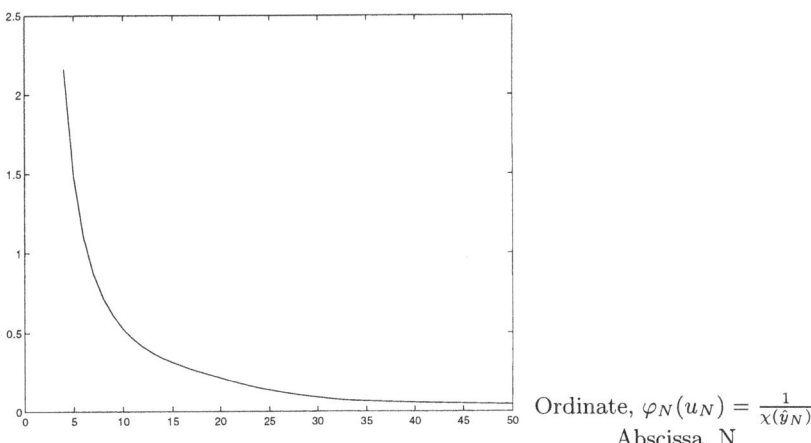

Ordinate, $\varphi_N(u_N) = \frac{1}{\chi(\hat{y}_N)}$
Abscissa, N

b) A Controlled Predator - Prey - Model:

We pick up the predator - prey - model that has been discussed in *Section 1.1.6 a)* whose controlled version we assume to be of the form

$$x_1(t+1) = x_1(t) + a\,x_1(t) - b\,x_1(t)x_2(t) - x_1(t)u_1(t) \ ,$$
$$x_2(t+1) = x_2(t) - c\,x_2(t) + d\,x_1(t)x_2(t) - x_2(t)u_2(t) \ , \ t \in \mathbb{N}_0 \ ,$$

where $a > 0$, $0 < c < 1$, $b > 0$, $d > 0$, $x_1(t)$ and $x_2(t)$ denote the density of the *prey* and *predator population* at time t, respectively, and $u_1, u_2 : \mathbb{N}_0 \to \mathbb{R}$ are control functions.

If we define

$$\tilde{g}_0(x_1(t), x_2(t)) = \begin{pmatrix} ax_1(t) - bx_1(t)x_2(t) \\ -cx_1(t) + dx_1(t)x_3(t) \end{pmatrix} \ , \ g_1(x_1(t), x_2(t)) = \begin{pmatrix} -x_1(t) \\ 0 \end{pmatrix}$$

and

$$g_2(x_1(t), x_2(t)) = \begin{pmatrix} 0 \\ -x_2(t) \end{pmatrix} \ ,$$

then the system can be rewritten in the form

$$x(t+1) = x(t) + \tilde{g}_0(x(t)) + \sum_{j=1}^{2} g_j(x(t))u_j(t) \ , \ t \in \mathbb{N}_0 \ , \tag{2.33}$$

with $x(t) = (x_1(t), x_2(t))^T$. This is exactly the system (2.31) for $m = 2$. In addition we assume an initial condition

$$x(0) = \begin{pmatrix} x_{01} \\ x_{02} \end{pmatrix} = x_0. \tag{2.34}$$

The uncontrolled system

$$x(t+1) = x(t) + \tilde{g}_0(x(t)) \ , \ t \in \mathbb{N}_0 \ ,$$

has $\hat{x} = \left(\frac{c}{d}, \frac{a}{b}\right)^T$ as fixed point.

We assume the set U of admissible control functions to be given by

$$U = \{u : \mathbb{N}_0 \to \mathbb{R}^2 \mid \|u(t)\|_2 \le \gamma \text{ for all } t \in \mathbb{N}_0\}$$

where $\gamma > 0$ is a given constant and $\|\cdot\|_2$ is the Euclidean norm. For every $u \in U$ we denote the unique solution $x : \mathbb{N}_0 \to \mathbb{R}^2$ of (2.33) and (2.34) by $x(u)$. Our aim now consists of finding some $N \in \mathbb{N}_0$ and a control function $u \in U$ with

$$u(t) = \Theta_2 \text{ for all } t \ge N \tag{2.35}$$

such that the solution $x : \mathbb{N}_0 \to \mathbb{R}^2$ of (2.33), (2.34) satisfies the end condition

$$x(N) = \hat{x}. \tag{2.36}$$

In order to find a solution of this problem we apply the iteration method described in *Section 2.1.3*. In the $k - th$ step of this procedure we have, for a given $u^{k-1} \in U$, to find some $N_k \in \mathbb{N}_0$ and a $u^k \in U$ with

$$u^k(t) = \Theta_2 \text{ for } t \geq N_k$$

such that

$$\sum_{t=1}^{N_k} \begin{pmatrix} x_1(u^{k-1})(t-1) & 0 \\ 0 & x_2(u^{k-1})(t-1) \end{pmatrix} \begin{pmatrix} u_1^k(t-1) \\ u_2^k(t-1) \end{pmatrix} = x^k - \hat{x},$$

where

$$x^k = x_0 + \sum_{t=1}^{N_k} \tilde{g}_0(x(u^{k-1})(t-1)) ,$$

and

$$\|u^k(t-1)\|_2 \leq \gamma \text{ for } t = 1, \dots, N_k.$$

For this we can apply the method developed in *Section 2.1.3*.

b) Control of a Planar Pendulum with Moving Suspension Point

We consider a non-linear planar pendulum of length l (> 0) whose movement is controlled by moving its suspension point with acceleration $u = u(t)$ along a horizontal straight line. If we denote the deviation angle from the orthogonal position of the pendulum by $\varphi = \varphi(t)$, then the movement of the pendulum is governed by the differential equation

$$\ddot{\varphi}(t) = -\frac{g}{l} \sin \varphi(t) - \frac{u(t)}{l} \cos \varphi(t) , \ t \in \mathbb{R} , \tag{2.37}$$

where g denotes the gravity constant.
For $t = 0$ initial conditions are given by

$$\varphi(0) = \varphi_0 \text{ and } \dot{\varphi}(0) = \dot{\varphi}_0 .$$

Now we discretize the differential equation by introducing a time step length $h > 0$ and replacing the second derivative $\ddot{\varphi}(t)$ by $\frac{1}{h^2}(\varphi(t+2h) - 2\varphi(t+h) + \varphi(t))$ thus obtaining the difference equation

$$\varphi(t+2h) = 2\varphi(t+h) - \varphi(t) - \frac{gh^2}{l} \sin \varphi(t) - \frac{u(t)h^2}{l} \cos \varphi(t) , \ t \in \mathbb{R}.$$

If we define

$$y_1(t) = \varphi(t) \text{ and } y_2(t) = \varphi(t+h) ,$$

then we obtain

$$y_1(t+h) = y_2(t) ,$$

$$y_2(t+h) = 2y_2(t) - y_1(t) - \frac{gh^2}{l} \sin y_1(t) - \frac{u(t)h^2}{l} \cos y_1(t) , \quad t \in \mathbb{R}.$$

Finally we define functions $x_1 : \mathbb{N}_0 \to \mathbb{R}$ and $x_2 : \mathbb{N}_0 \to \mathbb{R}$ by putting

$$x_1(n) = y_1(n \cdot h) \quad \text{and} \quad x_2(n) = y_2(n \cdot h) , \quad n \in \mathbb{N}_0 ,$$

and obtain the system

$$x_1(t+1) = x_2(t) , \tag{2.1'}$$

$$x_2(t+1) = 2x_2(t) - x_1(t) - \frac{gh^2}{l} \sin x_1(t) - \frac{u(t)h^2}{l} \cos x_1(t) ,$$

$t \in \mathbb{N}_0$ which is of the form (2.31) for $m = 1$ with

$$\tilde{g}_0(x_1(t), x_2(t)) = \begin{pmatrix} x_2(t) - x_1(t) \\ x_2(t) - x_1(t) - \frac{gh^2}{l} \sin x_1(t) \end{pmatrix} ,$$

$$g_1(x_1(t), x_2(t)) = \begin{pmatrix} 0 \\ -\frac{h^2}{l} \cos x_1(t) \end{pmatrix} , \quad t \in \mathbb{N}_0.$$

In addition we assume initial conditions

$$x_1(0) = \varphi_0 \quad \text{and} \quad x_2(0) = \dot{\varphi}_0. \tag{2.2'}$$

The uncontrolled system

$$x_1(t+1) = x_2(t) ,$$

$$x_2(t+1) = 2x_2(t) - x_1(t) - \frac{gh^2}{l} \sin x_1(t)$$

has $\hat{x} = (0,0)$ as fixed point.

We assume the set U of admissible control functions to be given by

$$U = \{u : \mathbb{N}_0 \to \mathbb{R} \mid |u(t)| \le \gamma \text{ for all } t \in \mathbb{N}_0\}$$

where $\gamma > 0$ is a given constant.

For every $u \in U$ we denote the unique solution $x : \mathbb{N}_0 \to \mathbb{R}^2$ of (2.1') and (2.2') by $x(u)$.

Our aim consists of finding some $N \in \mathbb{N}_0$ and a control function $u \in U$ with

$$u(t) = 0 \quad \text{for all } t \ge N \tag{2.7'}$$

such that the solution $x : \mathbb{N}_0 \to \mathbb{R}^2$ of (2.1'), (2.2') satisfies the end condition

$$x(N) = \Theta_2 .$$

This condition is equivalent to

$$x_2(N) = x_2(N-1) = 0.$$

The $k - th$ step of the iteration method described in *Section 2.1.3* for the solution of this problem then reads as follows: Let $u^{k-1} \in U$ be given. Then we determine $N^k \in \mathbb{N}_0$ and $u^k \in U$ such that

$$u^k(t) = 0 \quad \text{for all} \quad t \geq N_k$$

and the solution $x(u^k) : \mathbb{N}_0 \to \mathbb{R}^n$ of *(2.2')* and

$$x_1(u^1(t+1)) = x_1(u^k)(t) + x_2(u^{k-1})(t) - x_1(u^{k-1})(t) ,$$
$$x_2(u^1(t+1)) = x_2(u^k)(t) + x_2(u^{k-1})(t) - x_1(u^{k-1})(t)$$
$$- \frac{gh^2}{l} \sin x_1(u^{k-1})(t) - \frac{h^2}{l} \cos x_1(u^{k-1})(t) u^k(t) ,$$

$t \in \mathbb{N}_0$, satisfies the end conditions

$$x_2(u^k)(N_k) = x_2(u^k)(N_k - 1) = 0.$$

These are equivalent to

$$-\tfrac{1}{l} \sum_{t=1}^{N_k-1} \cos x_1(u^{k-1})(t-1) u^k(t-1) =$$

$$-\dot{\varphi}_0 - x_2(u^{k-1})(N_k - 2) + \varphi_0 - \tfrac{g}{l} \sum_{t=1}^{N_k-1} \sin x_1(u^{k-1})(t-1)$$

$$-\tfrac{1}{l} \cos x_1(u^{k-1})(N_k - 1) u^k(N_k - 1) =$$
$$-x_2(u^{k-1})(N_k - 1) + x_2(u^{k-1})(N_k - 2) - \tfrac{g}{l} \sin x_1(u^{k-1})(N_k - 1)$$

Let us consider the special case $N_k = 2$. Then it follows that

$$-\frac{1}{l} \cos \varphi_0 u^k(0) = \varphi_0 - \frac{g}{l} \sin \varphi_0 ,$$

$$-\frac{1}{l} \cos \dot{\varphi}_0 u^k(1) = -\dot{\varphi}_0 + \varphi_0 + \frac{gh^2}{l} \sin \varphi_0 - \frac{g}{l} \sin \dot{\varphi}_0 - \frac{h^2}{l} \cos \dot{\varphi}_0 u^{k-1}(0) .$$

Let us assume that
$$\cos \varphi_0 \neq 0 \quad \text{and} \quad \cos \dot{\varphi}_0 \neq 0 .$$

Then it follows for all $k \geq 1$ that

$$u^k(0) = \frac{-l\varphi_0}{\cos \varphi_0} + g \tan \varphi_0$$

and

$$u^k(1) = \frac{l(\varphi_0 - \dot{\varphi}_0)}{\cos \dot{\varphi}_0} - gh^2 \frac{\sin \varphi_0}{\cos \dot{\varphi}_0} + g \tan \dot{\varphi}_0 + h^2 u^{k-1}(0)$$

2.2 The Non-Autonomous Case

2.2.1 The Problem of Fixed Point Controllability

We consider a system of difference equations of the form

$$x(t+1) = g_t(x(t), u(t)) \ , \ t \in \mathbb{N}_0 \ , \tag{2.38}$$

where $(g_t)_{t \in \mathbb{N}_0}$ is a sequence of continuous vector functions $g_t : \mathbb{R}^n \times \mathbb{R}^m \to \mathbb{R}^n$, $u : \mathbb{N}_0 \to \mathbb{R}^m$ is a given vector function which is called a control function. The vector function $x : \mathbb{N}_0 \to \mathbb{R}^n$ which is called a state function is uniquely defined by (2.38), if we require an initial condition

$$x(0) = x_0 \tag{2.39}$$

for some given vector $x_0 \in \mathbb{R}^n$.
If we fix the control function $u : \mathbb{N}_0 \to \mathbb{R}^m$ and define

$$f_t(x) = g_t(x, u(t)) \ , \ x \in \mathbb{R}^n \ , \ t \in \mathbb{N}_0 \ , \tag{2.40}$$

then, for every $t \in \mathbb{N}_0$, $f_t : \mathbb{R}^n \to \mathbb{R}^n$ is a continuous mapping and $(\mathbb{R}^n, (f_t)_{t \in \mathbb{N}_0})$ is a *non-autonomous* time-discrete dynamical system which is controlled by the function $u : \mathbb{N}_0 \to \mathbb{R}^m$. If

$$u(t) = \Theta_m \ \text{ for all } \ t \in \mathbb{N}_0 \ ,$$

then the system (2.40) is called *uncontrolled*. Let us assume that the uncontrolled system (2.40) admits a fixed point $\hat{x} \in \mathbb{R}^n$ which then solves the equations

$$\hat{x} = g_t(\hat{x}, \Theta_m) \ \text{ for all } \ t \in \mathbb{N}. \tag{2.41}$$

Now let $\Omega \subseteq \mathbb{R}^m$ be a subset with $\Theta_m \in \Omega$. Then we define the set of admissible control functions by

$$U = \{u : \mathbb{N}_0 \to \mathbb{R}^m \mid u(t) \in \Omega \ \text{ for all } \ t \in \mathbb{N}_0\} \ . \tag{2.42}$$

After these preparations we can formulate the

Problem of Local Fixed Point Controllability

Given a fixed point $\hat{x} \in \mathbb{R}^n$ of the system

$$x(t+1) = g_t(x(t), \Theta_m) \ , t \in \mathbb{N}_0 \ , \tag{2.43}$$

i.e. a solution $\hat{x}$ of the equations (2.41) and an initial state $x_0 \in \mathbb{R}^n$, find some $N \in \mathbb{N}_0$ and a control function $u \in U$ with

$$u(t) = \Theta_m \quad \text{for all } t \geq N \tag{2.44}$$

such that the solution $x : \mathbb{N}_0 \to \mathbb{R}^n$ of (2.38), (2.39) satisfies the end condition

$$x(N) = \hat{x} \tag{2.45}$$

(which implies $x(t) = \hat{x}$ for all $t \geq N$). Let us assume that, for every $t \in \mathbb{N}_0$, the Jacobi-matrices

$$A_t = \frac{\partial g_t}{\partial x}(\hat{x}, \Theta_m) \quad \text{and} \quad B_t = \frac{\partial g_t}{\partial x}(\hat{x}, \Theta_m)$$

exist. Then it follows that, for every $t \in \mathbb{N}_0$,

$$x(t+1) - \hat{x} = g_t(x(t), u(t)) - g_t(\hat{x}, \Theta_m)$$
$$\approx A_t(x(t) - \hat{x}) + B_t u(t).$$

Therefore we replace the system (2.38) by

$$h(t+1) = A_t h(t) + B_t u(t) , \quad t \in \mathbb{N}_0 , \tag{2.46}$$

and the initial condition (2.39) by

$$h(0) = x_0 - \hat{x} . \tag{2.47}$$

The end condition (2.45) is replaced by

$$h(N) = \Theta_n = \text{ zero vector of } \mathbb{R}^n. \tag{2.48}$$

Now we consider the

Problem of Local Controllability

Find $N \in \mathbb{N}_0$ and $u \in U$ with

$$u(t) = \Theta_m \text{ for all } t \geq N$$

such that the corresponding solution $h : \mathbb{N}_0 \to \mathbb{R}^n$ of (2.46) and (2.47) satisfies the end condition (2.48) which implies

$$h(t) = \Theta_n \text{ for all } t \geq N .$$

From (2.46) and (2.47) we conclude that, for every $N \in \mathbb{N}$,

$$h(N) = A_{N-1} \ldots A_0(x_0 - \hat{x}) + \sum_{k=1}^{N} A_{N-1} \ldots A_k B_{k-1} u(k-1) ,$$

if, for $k = N$, we put $A_{N-1} \ldots A_k = I = n \times n$-unit matrix.

Therefore the end condition (2.48) is equivalent to

$$\sum_{k=1}^{N} A_{N-1}A_{N-2}\ldots A_k B_{k-1}u(k-1) = -A_{N-1}A_{N-2}\ldots A_0(x-\hat{x}) . \quad (2.49)$$

Let us assume that, for some $N_0 \in \mathbb{N}$,

$$rank(B_{N_0-1} \mid A_{N_0-1}B_{N_0-2} \mid \ldots \mid A_{N_0-1}\ldots A_1 B_0) = n . \quad (2.50)$$

Then, for $N = N_0$, the system (2.49) has a solution $(u(0)^T, u(1)^T, \ldots,$ $u(N_0-1)^T)^T \in \mathbb{R}^{m \cdot N_0}$ where $x_0 \in \mathbb{R}^N$ can be chosen arbitrarily. If we define

$$u(t) = \Theta_m \text{ for all } t \geq N_0 , \quad (2.51)$$

then we obtain a control function $u : \mathbb{N}_0 \to \mathbb{R}^m$ such that the corresponding solution $h : \mathbb{N}_0 \to \mathbb{R}^n$ of (2.46) and (2.47) satisfies the end condition (2.48) for $N = N_0$.
If A_k is non-singular for all $k \in \mathbb{N}_0$, then the assumption (2.50) implies that (2.50) holds true for all $N \geq N_0$ instead of N_0.
For instance, if we replace N_0 by $N_0 + 1$, then

$$(B_{N_0} \mid A_{N_0}B_{N_0-1} \mid A_{N_0}A_{N_0-1}B_{N_0-2} \mid \ldots \mid A_{N_0}\ldots A_1 B_0)$$

$$= (B_{N_0} \mid \underbrace{A_{N_0}}_{non-singular} \underbrace{(B_{N_0-1} \mid A_{N_0-1}B_{N_0-2} \mid \ldots \mid A_{N_0-1}\ldots A_1 B_0))}_{rank = n} .$$

$$\Rightarrow rank = n$$

So the system (2.49) has a solution $(u(0)^T, u(1)^T, \ldots, u(N_0-1)^T)^T$ for all $N \geq N_0$.
Next we assume that $\Omega \subseteq \mathbb{R}^m$ is convex, has Θ_m as interior point and satisfies $u \in \Omega \Rightarrow -u \in \Omega$. Let us define, for every $N \in \mathbb{N}$, the set

$$R(N) = \{ x = \sum_{k=1}^{N} A_{N-1}A_{N-2}\ldots A_k B_{k-1}u(k-1) \mid u \in U \} .$$

Then we can prove (see *Theorem 2.1*)

Theorem 2.5. *If, for some $N_0 \in \mathbb{N}$, condition (2.50) is satisfied and if A_k is non-singular for all $k \in \mathbb{N}_0$, then Θ_n is an interior point of $R(N)$ for all $N \geq N_0$.*

Proof. Let us assume that $\Theta_n (\in R(N))$ is not an interior point of $R(N)$ for some $N \geq N_0$. Thus $R(N)$ must be contained in a hyperplane through Θ_n, i.e., there must exist some $y \in \mathbb{R}^n$, $y \neq \Theta_n$, with

$$y^T x = 0 \text{ for all } x \in R(N).$$

This implies

$$\sum_{k=1}^{N} y^T A_{N-1} A_{N-2} \dots A_k B_{k-1} u^k = 0 \text{ for all } (u^{1^T}, \dots, u^{N^T})^T \in (\mathbb{R}^m)^N ,$$

hence

$$y^T A_{N-1} A_{N-2} \dots A_k B_{k-1} = \Theta_m^T \text{ for all } k = 1, \dots, N.$$

Since (2.50) also holds true for all $N \geq N_0$ instead of N_0, it follows that $y = \Theta_n$ which contradicts $y \neq \Theta_n$. Hence the assumption is false and the proof is complete.

□

As a consequence of *Theorem 2.5* we obtain

Theorem 2.6. *In addition to the assumption of Theorem 2.5 let*

$$\sup_{k \in \mathbb{N}_0} \|A_k\| < 1 \text{ where } \| \cdot \| \text{ denotes the spectral norm .} \qquad (2.52)$$

Then there is some $N \in \mathbb{N}$ and some $u \in U$ such that (2.49) holds true.

Proof. Assumption (2.52) implies

$$\lim_{N \to \infty} A_{N-1} A_{N-2} \dots A_0 (x_0 - \hat{x}) = \Theta_n .$$

Hence *Theorem 2.5* implies that there is some $N \in \mathbb{N}$ with $N \geq N_0$ such that

$$-A_{N-1} A_{N-2} \dots A_0 (x_0 - \hat{x}) \in R(N)$$

which completes the proof.

□

2.2.2 The General Problem of Controllability

We consider the same situation as at the beginning of *Section 2.2.1*. However, we do not assume the existence of a fixed point $\hat{x} \in \mathbb{R}^n$ of the uncontrolled system (2.43), i.e., a solution of (2.41). Instead we assume a vector $x_1 \in \mathbb{R}^n$ to be given and consider the general

Problem of Controllability

Find some $N \in \mathbb{N}_0$ and a control function $u \in U$ (2.42) such that the solution $x : \mathbb{N}_0 \to \mathbb{R}^n$ of (2.38), (2.39) satisfies the end condition

$$x(N) = x_1 . \qquad (2.53)$$

From (2.38) and (2.39) we infer

$$
\begin{aligned}
x(N) &= g_{N-1}(g_{N-2}(\ldots (g_0(x_0, u(0)), u(1)), \ldots), u(N-1)) \\
&= G^N(x_0, u(0), \ldots, u(N-1)).
\end{aligned}
\tag{2.54}
$$

Hence the end condition (2.53) is equivalent to

$$
G^N(x_0, u(0), \ldots, u(N-1)) = x_1
\tag{2.55}
$$

So we have to find vectors $u(0), \ldots, u(N-1) \in \Omega$ such that (2.55) is satisfied. For every $N \in \mathbb{N}$ we define the controllable set

$$
\begin{aligned}
S_N(x_1) = \{ \, x \in \mathbb{R}^n \mid &\text{ there exists some } u \in U \\
&\text{ such that } G^N(x, u(0), \ldots, u(N-1)) = x_1 \, \}
\end{aligned}
$$

and put

$$
S(x_1) = \bigcup_{N \in \mathbb{N}} S_N(x_1).
$$

Now let $x_0 \in S(x_1)$. Then we ask the question under which conditions is x_0 an interior point of $S(x_1)$?
In order to find an answer to this question we assume that

$$
\Omega \text{ is open and } g_N \in C^1(\mathbb{R}^n \times \mathbb{R}^m) \text{ for all } N \in \mathbb{N}_0 \, .
$$

Then it follows that $G^N \in C^1(\mathbb{R}^n \times \mathbb{R}^{m \cdot N})$ for every $N \in \mathbb{N}$ and

$$
\begin{aligned}
G_x^N(x, u(0), \ldots, u(N-1)) =& \\
(g_{N-1})_x(G^{N-1}(x, u(0), \ldots, u(N-2)), u(N-1)) \times& \\
(g_{N-2})_x(G^{N-2}(x, u(0), \ldots, u(N-3)), u(N-2)) \times& \\
\vdots& \\
(g_0)_x(x, u(0)) \, ,&
\end{aligned}
$$

and

$$
\begin{aligned}
G_{u(k)}^N(x, u(0), \ldots, u(N-1)) =& \\
(g_{N-1})_x(G^{N-1}(x, u(0), \ldots, u(N-2)), u(N-1)) \times& \\
(g_{N-2})_x(G^{N-2}(x, u(0), \ldots, u(N-3)), u(N-2)) \times& \\
\vdots& \\
(g_{k+1})_x(G^{k+1}(x, u(0), \ldots, u(k)), u(k+1)) \times& \\
(g_k)_{u(k)}(G^k(x, u(0), \ldots, u(k-1)), u(k))&
\end{aligned}
$$

for $k = 0, \ldots, N-1$, $x \in \mathbb{R}^n$ and $u \in U$.

Let us assume that $x_0 \in S_{N_0}(x_1)$ for some $N_0 \in \mathbb{N}$, i.e.

$$G^{N_0}(x_0, u_0(0), \ldots, u_0(N_0 - 1)) = x_1$$

for some $u_0 \in U$. Further let $(g_N)_x(x, u_0)$ be non-singular for all $N \in \mathbb{N}_0$, for all $x \in \mathbb{R}^n$ and all $u \in \Omega$.

Then $G_x^{N_0}(x_0, u_0(0), \ldots, u_0(N_0 - 1))$ is also non-singular and, by the implicit function theorem, there exists an open set $V \subseteq \Omega^{N_0}$ with $(u_0(0), \ldots, u_0(N_0 - 1)) \in V$ and a function $h : V \to \mathbb{R}^n$ with $h \in C^1(V)$ such that

$$h(u_0(0)), \ldots, u_0(N_0 - 1)) = x_0$$

and

$$G^{N_0}(h(u(0), \ldots, u(N_0 - 1)), u(0), \ldots, u(N_0 - 1)) = x_1$$
$$\text{for all } (u(0), \ldots, u(N_0 - 1)) \in V$$

which means

$$h(u(0), \ldots, u(N_0 - 1)) \in S_{N_0}(x_1)$$
$$\text{for all } (u(0), \ldots, u(N_0 - 1)) \in V .$$

Moreover,

$$h_{u(k)}(u_0(0), \ldots, u_0(N_0 - 1)) = -G_x^{N_0}(x_0, u_0(0), \ldots, u_0(N_0 - 1))^{-1} \times$$
$$G_{u(k)}^{N_0}(x_0, u_0(0), \ldots, u_0(N_0 - 1)) .$$

Next we assume that

$$rank(h_{u(0)}(u_0(0), \ldots, u_0(N_0-1)) \mid \ldots \mid h_{u(N_0-1)}(u_0(0), \ldots, u_0(N_0-1))) = n .$$

Then it follows with the aid of the inverse function theorem that there exists an n-dimensional relatively open set $\tilde{V} \subseteq V$ with $(u_0(0), \ldots, u_0(N_0 - 1)) \in \tilde{V}$ such that the restriction of h to $\tilde{V}$ is a homeomorphism which implies that $h(\tilde{V}) \subseteq S_{N_0}(x_1)$ is open.
Therefore $x_0 \in h(\tilde{V})$ is an interior point of $S(x_1)$.

Now we consider the special case where there exists some $\hat{x} \in \mathbb{R}^n$ with

$$g_N(\hat{x}, \Theta_m) = \hat{x} \text{ for all } N \in \mathbb{N}$$

which implies

$$G^N(\hat{x}, \Theta_m^N) = \hat{x} \text{ for all } N \in \mathbb{N} .$$

Then, for every $N \in \mathbb{N}$, it follows that $\hat{x} \in S_N(\hat{x})$, hence $\hat{x} \in S(\hat{x})$.
Let us assume that

$$(g_N)_x(\hat{x}, \Theta_m) \text{ is non-singular for all } N \in \mathbb{N}_0 .$$

Then

$$G_x^N(\hat{x}, \Theta_m^N) = (g_{N-1})_x(\hat{x}, \Theta_m) \cdot (g_{N-2})_x(\hat{x}, \Theta_m) \cdots (g_0)_x(\hat{x}, \Theta_m)$$

is also non-singular for all $N \in \mathbb{N}$.

By the implicit function theorem we therefore conclude, for every $N \in \mathbb{N}$, that there exists an open set $V_N \subseteq \Omega^N$ with $\Theta_m^N \in V_N$ and a function $h_N : V_N \to \mathbb{R}^n$ with $h_N \in C^1(V_N)$ such that

$$h_N(\Theta_m^N) = \hat{x} \text{ and } G^N(h_N(u(0), \ldots, u(N-1)), u(0), \ldots, u(N-1)) = \hat{x}$$

$$\text{for all } (u(0), \ldots, u(N-1)) \in V_N$$

which means

$$h_N(u(0), \ldots, u(N-1)) \in S_N(\hat{x})$$

$$\text{for all } (u(0), \ldots, u(N-1)) \in V_N .$$

Moreover,

$$(h_N)_{u(k)}(\Theta_m^N) = -G_x^N(\hat{x}, \Theta_m^N)^{-1} \cdot G_{u(k)}^N(\hat{x}, \Theta_m^N) .$$

Next we assume that, for some $N_0 \in \mathbb{N}$,

$$rank((h_{N_0})_{u(0)}(\Theta_m^{N_0}) \mid \cdots \mid (h_{N_0})_{u(N_0-1)}(\Theta_m^{N_0})) = n .$$

Then it follows with the aid of the inverse function theorem that there exists an n-dimensional relatively open set $\tilde{V}_{N_0} \subseteq V_{N_0}$ with $\Theta_m^{N_0} \in \tilde{V}_{N_0}$ such that the restriction of h_{N_0} to $\tilde{V}_{N_0}$ is a homeomorphism which implies that $h_{N_0}(\tilde{V}_{N_0}) \subseteq S_{N_0}(\hat{x})$ is open. Therefore $\hat{x} \in h_{N_0}(\tilde{V}_{N_0})$ is an interior point of $S(\hat{x})$. This result is a generalization of *Theorem 2.5*, if Ω in addition is open.

2.2.3 Stabilization of Controlled Systems

Let $(g_t)_{t \in \mathbb{N}}$ be a sequence of continuous mappings $g_t : \mathbb{R}^n \times \mathbb{R}^m \to \mathbb{R}^n$ and let $\mathcal{H}$ be a family of continuous mappings $h : \mathbb{R}^n \to \mathbb{R}^m$. If we define, for every $h \in \mathcal{H}$ and $t \in \mathbb{N}$, the mapping $f_t^h : \mathbb{R}^n \to \mathbb{R}^n$ by

$$f_t^h(x) = g_t(x, h(x)) , \quad x \in \mathbb{R} ,$$

then we obtain a non-autonomous time-discrete dynamical system $(\mathbb{R}^n, (f_t^h)_{t \in \mathbb{N}})$. The dynamics in this system is defined by the sequence $F^h = (F_t^h)_{t \in \mathbb{N}}$ of mappings $F_t^h : \mathbb{R}^n \to \mathbb{R}^n$ given by

$$F_t^h(x) = f_t^h \circ f_{t-1}^h \circ \cdots \circ f_1^h(x) \text{ for all } x \in \mathbb{R}^n \text{ and } t \in \mathbb{N}$$

and

$$F_0^h(x) = x \text{ for all } x \in \mathbb{R}^n .$$

We also obtain the dynamical system $(\mathbb{R}^n, (f_t^h)_{t \in \mathbb{N}})$, if we replace the control function $u : \mathbb{N}_0 \to \mathbb{R}^m$ in the system (2.38) by the feedback controls $h(x) : \mathbb{N}_0 \to \mathbb{R}^m, x \in \mathbb{R}^n$.

The problem of stabilization of the controlled system (2.38) by the feedback controls $h(x), x \in \mathbb{R}^n$, then reads as follows: Given $x_0 \in \mathbb{R}^n$ such that the limit set $L_{F^h}(x_0)$ defined by (1.6') (see *Section 1.2.1*) is non-empty and compact for all $h \in \mathcal{H}$.

Find a mapping $h \in \mathcal{H}$ such that $L_{F^h}(x_0)$ is stable, an attractor or asymptotically stable with respect to $(f_t^h)_{t \in \mathbb{N}}$.

Let us consider the special case

$$g_t(x, u) = A_t(x)x + B_t(x)u \text{ for } x \in \mathbb{R}^n , \ u \in \mathbb{R}^m , \qquad (2.56)$$

where $(A_t(x))_{t \in \mathbb{N}}$ and $(B_t(x))_{t \in \mathbb{N}}$ is a sequence of real, continuous $n \times n$- and $n \times m$ matrix functions on $\mathbb{R}^n$, respectively.

Let $\mathcal{H}$ be the family of all linear mappings $h : \mathbb{R}^n \to \mathbb{R}^m$ (which are automatically continuous). Every $h \in \mathcal{H}$ is then representable in the form

$$h(x) = C^h x , \quad x \in \mathbb{R}^n ,$$

where C^h is a real $m \times n$-matrix. For every $t \in \mathbb{N}$ and $h \in \mathcal{H}$ we therefore obtain

$$f_t^h(x) = (A_t(x) + B_t(x)C^h)x , \ x \in \mathbb{R}^n . \qquad (2.57)$$

Let us put

$$D_t h(x) = A_t(x) + B_t(x)C^h \text{ for } x \in \mathbb{R}^n .$$

If we choose $x_0 = \Theta_n =$ zero vector of $\mathbb{R}^n$, then we conclude

$$F_t^h(x_0) = x_0 \text{ for all } t \in \mathbb{N}_0 , \ h \in \mathcal{H} ,$$

and therefore $L_{F^h}(x_0) = \{x_0\}$.

The problem of stabilization of the controlled system (2.38) with g_t , $t \in \mathbb{N}$, given by (2.56) in this situation consists of finding an $m \times n$-matrix C^h such that $\{x_0 = \Theta_n\}$ is stable, an attractor or asymptotically stable with respect to $(f_t^h)_{t \in \mathbb{N}}$ with f_t^h given by (2.57).

Now let us assume that

$$\|D_t^h(x)\| \le 1 \text{ for all } x \in \mathbb{R}^n \text{ and } t \in \mathbb{N} \qquad (2.58)$$

where $\| \cdot \|$ denotes the spectral norm.

Let $U \subseteq \mathbb{R}^n$ be a relatively compact open set with $x_0 = \Theta_n \in U$.

Then there is some $r > 0$ such that

$$B_U = \{x \in \mathbb{R}^n \mid \|x\|_2 < r\} \subseteq U .$$

Hence B_U is open, $x_0 \in B_U$, and assumption (2.58) implies

$$f_t^h(B_u) \subseteq B_U \text{ for all } t \in \mathbb{N} .$$

If we define

$$V(x) = \|x\|_2^2 = x^T x \text{ for } x \in \mathbb{R}^n ,$$

then

$$V(x) \geq 0 \text{ for all } x \in \mathbb{R}^n \text{ and } (V(x) = 0 \Leftrightarrow x = x_0 = \Theta_n)$$

and

$$\begin{aligned}
V(f_t^h(x)) - V(x) &= x^T D_t^h(x)^T D_t^h(x)x - x^T x \\
&= \|D_t^h(x)x\|_2^2 - \|x\|_2^2 \leq (\|D_t^h(x)\| - 1)\|x\|_2^2 \leq 0
\end{aligned}$$

for all $x \in \mathbb{R}^n$ and $t \in \mathbb{N}$.

This shows that V is a Lyapunov function with respect to $(f_t^h)_{t \in \mathbb{N}}$ on $G = \mathbb{R}^n$ which is positive definite with respect to $\{x_0 = \Theta_n\}$. By *Theorem 1.1'* we therefore conclude that $\{x_0 = \Theta_n\}$ is stable with respect to $(f_t^h)_{t \in \mathbb{N}}$ with f_t^h given by (2.57).

Next we assume that

$$\sup_{t \in \mathbb{N}} \|D_t^h(x)\| < 1 \text{ for all } x \in \mathbb{R}^n . \tag{2.59}$$

Then it follows from

$$\begin{aligned}
V(F_t^h(x)) &= x^T D_1^h(x)^T \dots D_t^h(F_{t-1}^h(x))^T D_t^h(F_{t-1}^h(x)) \dots D_1^h(x)x \\
&= \|D_t^h(F_{t-1}^h(x)) \dots D_1^h(x)x\|_2^2 \leq \|D_t^h(F_{t-1}^h(x))\|^2 \dots \|D_1^h(x)\|^2 \|x\|^2
\end{aligned}$$

for all $x \in \mathbb{R}^n$ and $t \in \mathbb{N}$ that

$$\lim_{t \to \infty} V(F_t^h(x)) = 0 \text{ for all } x \in \mathbb{R}^n .$$

This implies

$$\lim_{t \to \infty} F_t^h(x) = \Theta_n \text{ for all } x \in \mathbb{R}^n$$

and shows that $\{x_0 = \Theta_n\}$ is an attractor with respect to $(f_t^h)_{t \in \mathbb{N}}$ with f_t^h given by (2.57).

Result. Under the assumption (2.59) the set $\{x_0 = \Theta_n\}$ is asymptotically stable with respect to $(f_t^h)_{t \in \mathbb{N}}$ with f_t^h given by (2.57).

2.2.4 The Problem of Reachability

We again consider the situation at the beginning of *Section 2.2.1* without necessarily assuming the existence of a fixed point $\hat{x} \in \mathbb{R}^n$ of the uncontrolled system (2.43). Let $\Omega \subseteq \mathbb{R}^m$ be a non-empty subset. For a given $x_0 \in \mathbb{R}^n$ we then define the set of states that are reachable from x_0 in $N \in \mathbb{N}$ steps by

$$R_N(x_0) = \{x = G^N(x_0, u(0), \ldots, u(N-1)) \mid u(k) \in \Omega , \; k = 0, \ldots, N-1\} \tag{2.60}$$

where the map $G^N : \mathbb{R}^{m \cdot N} \to \mathbb{R}^n$ is defined by (2.54).
Further we define the set of states reachable from x_0 in a suitable number of steps by

$$R(x_0) = \bigcup_{N \in \mathbb{N}} R_N(x_0) . \tag{2.61}$$

The question we are interested in now is: Under which conditions does $R(x_0)$ have a non-empty interior?
A simple answer to this question gives

Theorem 2.7. *Let Ω be open. If there is some $N \in \mathbb{N}$ and there exist $u(0), \ldots, u(N-1) \in \Omega$ such that*

$$rank(G^N_{u(0)}(x_0, u(0), \ldots, u(N-1)) \mid \ldots$$
$$\ldots \mid G^N_{u(N-1)}(x_0, u(0), \ldots, u(N-1))) = n , \tag{2.62}$$

then $R_N(x_0)$ has a non-empty interior and therefore also $R(x_0)$.

Proof. Condition (2.62) implies that the $n \times N \cdot m$-matrix

$$\left(G^N_{u(0)}(x_0, u(0), \ldots, u(N-1)) \mid \ldots \mid G^N_{u(N-1)}(x_0, u(0), \ldots, u(N-1)) \right)$$

has n linearly independent column vectors. Let E be the n-dimensional subset of Ω^N consisting of all vectors whose components which do not correspond to these linearly independent column vectors are equal to the ones of $(u(0)^T, \ldots, u(N-1)^T)^T$. If we restrict the mapping G^N to E, then the Jacobi matrix of this restriction consists of these linearly independent column vectors and is therefore non-singular.

By the inverse function theorem therefore there exists an open set (with respect to E) $U \subseteq \Omega^N$ with $\left((u(0)^T, \ldots, u(N-1)^T)\right)^T \in U$ which is mapped homeomorphically by G^N on an open set $V \subseteq R_N(x_0)$ with $G^N(x_0, u(0), \ldots, u(N-1)) \in V$. This completes the proof.

$\square$

Next let us consider the linear case where

$$g_t(x, u) = A_t x + B_t u \quad , \quad x \in \mathbb{R}^n, \ u \in \mathbb{R}^m,$$

with $n \times n-$ and $n \times m-$ matrices A_t and B_t, respectively, for every $t \in \mathbb{N}_0$. Then, for every $N \in \mathbb{N}$ and every $x_0 \in \mathbb{R}^n$, we obtain

$$G^N(x_0, u(0), \ldots, u(N-1)) = A_{N-1} \ldots A_0 x_0 + \sum_{k=1}^{N} A_{N-1} \ldots A_k B_{k-1} u(k-1)$$

where for $k = N$ we put $A_{N-1} \ldots, A_k = I = n \times n-$unit matrix. Further we have, for every $N \in \mathbb{N}$ and every $x_0 \in \mathbb{R}^n$,

$$R_N(x_0) = \{x = A_{N-1} \ldots A_0 x_0 + \sum_{k=1}^{N} A_{N-1} \ldots A_k B_{k-1} u(k-1) \mid$$
$$u(k) \in \Omega, \ k = 0, \ldots, N-1\}.$$

Because of

$$G^N_{u(k-1)}(x_0, u(0), \ldots, u(N-1)) = A_{N-1} \ldots A_k B_{k-1} \quad \text{for} \quad k = 1, \ldots, N$$

it follows that the condition (2.62) for $N = N_0$ coincides with the condition (2.50).
If this is satisfied, then by *Theorem 2.7* the set $R(x_0)$ (2.61) states reachable from x_0 has a non-empty interior.
If $\Omega = \mathbb{R}^m$, it follows in addition that $R(x_o) = \mathbb{R}^n$ for all $x_0 \in \mathbb{R}^n$.

Proof. Let $x, x_0 \in \mathbb{R}^n$ be given arbitrarily. Then condition (2.50) implies the existence of $u(k) \in \mathbb{R}^m$ for $k = 0, \ldots, N_0 - 1$ such that

$$x - A_{N_0-1} \ldots A_0 x_0 = \sum_{k=1}^{N_0} A_{N_0-1} \ldots A_k B_{k-1} u(k-1)$$

hold true which shows that $x \in \mathbb{R}_{N_0}(x_0) \subseteq R(x_0)$.

$\square$

For every $k = 1, \ldots, N$ let us define an $n \times m-$matrix C^k by

$$C^k = A_{N-1} \ldots A_k B_{k-1} \quad \text{for} \quad k = 1, \ldots, N-1$$

and

$$C^N = B_{N-1}.$$

The condition (2.50) implies the existence of n column vectors

$$\begin{pmatrix} c_{ij_{k_l}}^{k_l} \\ \vdots \\ c_{nj_{k_l}} \end{pmatrix} \quad \text{for} \quad l = 1, \ldots, n \quad \text{which are linearly independent .}$$

If we define the $n \times n-$matrix C and a vector $u \in \mathbb{R}^n$ by

$$C = \begin{pmatrix} c_{1j_{k_1}}^{k_1} & \cdots & c_{1j_{k_n}}^{k_n} \\ \vdots & & \vdots \\ c_{nj_{k_1}}^{k_1} & \cdots & c_{nj_{k_n}}^{k_n} \end{pmatrix} \quad \text{and} \quad u = \begin{pmatrix} u_{j_{k_1}}(k_1 - 1) \\ \vdots \\ u_{j_{k_n}}(k_n - 1) \end{pmatrix}, \text{ respectively,}$$

and put

$$u_j(k - 1) = 0 \quad \text{for} \quad k \neq k_l \ , \ j \neq j_{k_l} \ , \ l = 1, \ldots, n \ ,$$

then we obtain

$$G^N(x_0, u(0), \ldots, u(N - 1)) = A_{N-1} \ldots A_0 x_0 + Cu$$

which implies

$$u = C^{-1} \left(\underbrace{G^N(x_0, u(0), \ldots, u(N - 1))}_{=x} - A_{N-1} \ldots A_0 x_0 \right) .$$

Now let $E = \{u = (u(0), \ldots, u(N - 1)) \in \mathbb{R}^{m \cdot N} \mid u_j(k - 1) = 0 \text{ for } k \neq k_l \text{ and } j \neq j_{k_l} \ , \ l = 1, \ldots, n\}$.
Then $G^N(x_0, \cdot)$ is a linear isomorphism from E on $\mathbb{R}^n$.
Therefore

$$G^N(x_0, u) = x \quad \text{for some} \quad u \in E$$

implies

$$u = G^N(x_0, \cdot)^{-1}(x) \text{ and } G^N(x_0, u) = A_{N-1} \ldots A_0 x_0 + C \cdot G^N(x_0, \cdot)^{-1}(x) .$$

If all A_k, $k \in \mathbb{N}_0$, are invertible, it follows that

$$x_0 = A_0^{-1} \ldots A_{N-1}^{-1} x - A_0^{-1} \ldots A_{N-1}^{-1} C \cdot G^N(x_0, \cdot)^{-1}(x) .$$

In the nonlinear case we have the following situation:
If the condition (2.62) is satisfied, there exists an $n-$dimensional subset E of Ω^N and a set $U \subseteq E$ which is open with respect to E and contains $(u(0)^T, \ldots, u(N - 1)^T)^T$ and which is mapped homeomorphically on an open $V \subseteq R_N(x_0)$ by the restriction of $G^N(x_0, \cdot)$ to E. If

$$x = G^N(x_0, u(0), \ldots, u(N - 1)) ,$$

then

$$(u(0)^T, \ldots, u(N - 1)^T) = G^N(x_o, \cdot)^{-1}(x) .$$

If in addition $G_x^N(x_0, u(0), \ldots, u(N-1))$ is non-singular, then by the implicit function theorem there exists an open set $W \subseteq \Omega^N$ which contains $(u(0)^T, \ldots, u(N-1)^T)^T$ and a function $h : W \to \mathbb{R}^n$ with $h \in C^1(W)$ such that

$$h(u(0), \ldots, u(N-1)) = x_0$$

and

$$G^N(h(\tilde{u}(0), \ldots, \tilde{u}(N-1)), \tilde{u}(0), \ldots, \tilde{u}(N-1)) = x$$
$$\text{for all } (\tilde{u}(0)^T, \ldots, \tilde{u}(N-1)^T)^T \in W .$$

This implies

$$x_0 = h(G^n(x_0, \cdot)^{-1}(x)) .$$

Since

$$h_{u(k)}(u(0), \ldots, u(N-1)) =$$
$$-G_x(x_0, u(0), \ldots, u(N-1))^{-1} \times G_{u(k)}^{N_0}(x_0, u(0), \ldots, u(N-1))$$
$$\text{for } k = 0, \ldots, N-1 ,$$

it follows that

$$rank(h_{u(0)}(u(0), \ldots, u(N-1)) \mid \ldots \mid h_{u(N-1)}(u(0), \ldots, u(N-1))) = n$$

which implies that h maps $U \cap W$ homeomorphically onto an open set $\tilde{V} \subseteq R_N(x_0)$ which contains x. Therefore $h \circ G^N(x_0, \cdot)^{-1}$ maps $V \cap \tilde{V}$ homeomorphically on an open set $\tilde{\tilde{V}}$ which contains x_0 and is contained in

$$S_N(x) = \{\tilde{x} \in \mathbb{R}^n \mid \text{there exists some } \tilde{u} \in U \quad (2.5)$$
$$\text{with } G^N(\tilde{x}, \tilde{u}(0), \ldots, \tilde{u}(N-1)) = x\} .$$

Finally let us assume that (as in *Theorem 2.7*) Ω is open and (for the given $x_0 \in \mathbb{R}^n$) there exists some $N \in \mathbb{N}$ such that the condition (2.62) is satisfied for all $(u(0)^T, \ldots, u(N-1)^T)^T \in \Omega^N$. Then it follows from the proof of *Theorem 2.7* that every $x \in \mathbb{R}_N(x_0)$ is an interior point of $R_N(x_0)$, i.e., $R_N(x_0)$ is open.

This implies in the linear case with Ω being an open subset of $\mathbb{R}^m$ that $R_N(x_0)$ is open for every $x_0 \in \mathbb{R}^n$, if the condition (2.50) is satisfied.

3

Controllability and Optimization

3.1 The Control Problem

We consider n players P_i, $i = 1, \ldots, n$, that are involved in a so called dynamical game. We assume that every player is assigned a state vector function $x_i : \mathbb{N}_0 \to \mathbb{R}^{n_i}$ and has at his disposal a control vector function $u_i : \mathbb{N}_0 \to \mathbb{R}^{m_i}$ which are dynamically coupled by a system of difference equations

$$x_i(t+1) = g_i(x(t), u(t)) , \ t \in \mathbb{N}_0 ,$$
$$i = 1, \ldots, n , \tag{3.1}$$

where $x(t) = (x_1(t)^T, \ldots, x_n(t)^T)^T$, $u(t) = (u_1(t)^T, \ldots, u_n(t)^T)^T$, and $g_i \in C(\mathbb{R}^N \times \mathbb{R}^M, \mathbb{R}^{n_i})$, $i = 1, \ldots, n$, with $N = \sum_{i=1}^{n} n_i$ and $M = \sum_{i=1}^{n} m_i$.

If one puts $g = (g_1^T, \ldots, g_n^T)^T$, then (3.1) can be written in the form

$$x(t+1) = g(x(t), \ u(t)), \ t \in \mathbb{N}_0 , \tag{3.2}$$

where $g \in C(\mathbb{R}^N \times \mathbb{R}^M, \mathbb{R}^N)$.

Let, for every $i = 1, \ldots, n$, $U_i \subseteq \mathbb{R}^{m_i}$ be a control set with $\Theta_{m_i} \in U_i (\Theta_{m_i} =$ zero vector in $\mathbb{R}^{m_i}$). Then the system

$$x(t+1) = g(x(t), \Theta_M) , \ t \in \mathbb{N}_0 , \tag{3.3}$$

is called uncontrolled.

Assumption: The uncontrolled system (3.3) possesses a fixed point $\hat{x} \in \mathbb{R}^N$, i.e., a solution of the equation

$$g(\hat{x}, \Theta_M) = \hat{x} . \tag{3.4}$$

We assume that this fixed point is a desired state of the whole system for all players.

Therefore they try to find control functions $u_i : \mathbb{N}_0 \to \mathbb{R}^{m_i}$, $i = 1, \ldots, n$, with

$$u_i(t) \in U_i \text{ for all } t \in \mathbb{N}_0$$

such that the system (3.2) is driven from a given initial state $x_0 \in \mathbb{R}^N$ into $\hat{x}$ in finitely many time steps.

This leads to the following

Problem of Controllability

Given an initial state $x_0 \in \mathbb{R}^N$, find a time $T \in \mathbb{N}_0$ and a control function $u : \mathbb{N}_0 \to \mathbb{R}^M$ with

$$u(t) \in U = \prod_{i=1}^{n} U_i \text{ for all } t \in \mathbb{N}_0 \tag{3.5}$$

and

$$u(t) = \Theta_M \text{ for all } t \geq T \tag{3.6}$$

such that the solution $x : \mathbb{N}_0 \to \mathbb{R}^N$ of (3.2) which satisfies the initial condition

$$x(0) = x_0 \tag{3.7}$$

satisfies the end condition

$$x(T) = \hat{x} \tag{3.8}$$

which implies

$$x(t) = \hat{x} \text{ for all } t \geq T .$$

From (3.2) and (3.7) it follows that

$$x(T) = \underbrace{g(g(\ldots(g(x_0, u(0)), u(1)), \ldots), u(T-1))}_{T-times}$$
$$= G^T(x_0, u(0), \ldots, u(T-1)) .$$

Now let $T \in \mathbb{N}$ be given. If the $u(0), \ldots, u(T-1) \in U$ are solutions of the system

$$G^T(x_0, u(0), \ldots, u(T-1)) = \hat{x} \tag{3.9}$$

and if one defines

$$u(t) = \Theta_M \text{ for all } t \geq T ,$$

then one obtains a control function $u : \mathbb{N}_0 \to \mathbb{R}^M$ with (3.5) and (3.6) which solves the problem of controllability.

3.2 A Game Theoretical Solution

3.2.1 The Cooperative Case

For every player $P_i, i = 1, \ldots, n$, we define a cost function

$$\varphi_i^T(u(0), \ldots, u(T-1)) = \|G_i^T(x_0, u(0), \ldots, u(T-1)) - \hat{x}_i\|_2^2$$

which he wants to become zero for a suitable choice of $u(k) \in U$ for $k = 0, \ldots, T-1$ where $\| \cdot \|_2$ denotes the Euclidean norm in $\mathbb{R}^{n_i}$, $i = 1, \ldots, n$. With this we define a total cost function by virtue of

$$\varphi^T(u(0), \ldots, u(T-1)) = \sum_{i=1}^{n} \varphi_i^T(u(0), \ldots, u(T-1)) .$$

To solve system (3.9) cooperatively now consists of finding $u^T(0), \ldots, u^T(T-1) \in U$ such that

$$\varphi^T(u^T(0), \ldots, u^T(T-1)) \leq \varphi^T(u(0), \ldots, u(T-1))$$
$$\text{for all } u(0), \ldots, u(T-1) \in U .$$
(3.10)

Every solution $u^T(0), \ldots, u^T(T-1) \in U$ of the optimization problem (3.10) is a so called *Pareto optimum*, i.e., the following holds true:
For every $u(0), \ldots, u(T-1) \in U$ with

$$\varphi_i^T(u(0), \ldots, u(T-1)) \leq \varphi_i^T(u^T(0), \ldots, u^T(T-1))$$
$$\text{for all } i = 1, \ldots, n$$
(3.11)

it follows necessarily that

$$\varphi_i^T(u(0), \ldots, u(T-1)) = \varphi_i^T(u^T(0), \ldots, u^T(T-1))$$
$$\text{for all } i = 1, \ldots, n$$
(3.12)

By contraposition this means: For every $u(0), \ldots, u(T-1) \in U$ with

$$\varphi_{i_0}^T(u(0), \ldots, u(T-1)) < \varphi_{i_0}^T(u^T(0), \ldots, u^T(T-1))$$

for some $i_0 \in \{1, \ldots, n\}$ there exists $i_1 \in \{1, \ldots, n\}$ such that

$$\varphi_{i_1}^T(u(0), \ldots, u(T-1)) > \varphi_{i_1}^T(u^T(0), \ldots, u^T(T-1)) .$$

In words : There is no T-tupel $(u(0), \ldots, u(T-1)) \in U^T$ for which one player can improve his cost function value in comparison with that of the T-tupel $(u^T(0), \ldots, u^T(T-1)) \in U^T$ without another player having to deteriorate his cost function value.

The implication (3.11) $\Rightarrow$ (3.12) can be easily seen as follows: (3.11) implies $\varphi^T(u(0), \ldots, u(T-1)) \le \varphi^T(u^T(0), \ldots, u^T(T-1))$ hence $\varphi^T(u(0), \ldots, u(T-1)) = \varphi^T(u^T(0), \ldots, u^T(T-1))$ which is only possible, if (3.12) holds true.

If

$$\varphi^T(u^T(0), \ldots, u^T(T-1)) = 0 ,$$

then $u^T(0), \ldots, u^T(T-1) \in U$ is a solution of the system (3.9). Further it follows that

$$\varphi^{T+1}(u^{T+1}(0), \ldots, u^{T+1}(T)) \le \varphi^{T+1}(u^T(0), \ldots, u^T(T-1), \Theta_M)$$

so that in a procedure for solving the minimization problem (3.10) for $T+1$ instead of T the $(T+1)$-tupel $(u^T(0), \ldots, u^T(T-1), \Theta_M)$ could be used as starting solution. The existence of a solution of the problem (3.10) is ensured, if every U_i for $i = 1, \ldots, n$ is compact in $\mathbb{R}^{m_i}$. The evaluation of the total cost function φ^T can be very complicated. We therefore present an iteration method for the solution of problem (3.10) for which the function evaluations are less complicated. Now, we replace the system (3.2) by

$$x(t+1) = x(t) + f(x(t), u(t)) , \quad t \in \mathbb{N}_0$$

(for instance, by defining $f(x, u) = g(x, u) - x$). Then we choose, for some given $T \in \mathbb{N}$,

$$x^0(t+1) = x^0(t) + f(x^0(t), u^0(t))$$

$$\text{for } t = 0, \ldots, T-1$$

where

$$x^0(0) = x_0 .$$

Then we construct a sequence $(u^k(0), \ldots, u^k(T-1))_{k \in \mathbb{N}_0}$ in U^T and a sequence $(x^k(1), \ldots, x^k(T))_{k \in \mathbb{N}_0}$ in $R^{N \cdot T}$ and $x^k(0) = x_0$, are given, then we determine $(u^{k+1}(0), \ldots, u^{k+1}(T-1)) \in U^T$ such that for

$$\tilde{x}^{k+1}(t+1) = \tilde{x}^k(t) + f(x^k(t), u^{k+1}(t)) \text{for } t = 0, \ldots, T-1$$

with

$$\tilde{x}^{k+1}(0) = x_0$$

the function value

$$\varphi_k^T(u^{k+1}(0), \ldots, u^{k+1}(T-1)) = \sum_{i=1}^{n} \|\tilde{x}_i^{k+1}(T) - \hat{x}_i\|_2^2$$

becomes minimal.

Here

$$\tilde{x}^{k+1}(T) = x_0 + f(x_0, u^{k+1}(0)) + \sum_{t=1}^{T-1} f(x^k(t), u^{k+1}(t))$$

and hence

$$\varphi_k^T(u^{k+1}(0), \ldots, u^{k+1}(T-1))$$

$$= \sum_{i=1}^{n} \| \sum_{t=0}^{T-1} f_i(x^k(t), u^{k+1}(t)) + x_{0i} - \hat{x}_i \|_2^2 .$$

If $(u^{k+1}(0), \ldots, u^{k+1}(T-1)) \in U^T$ has been determined, then we define

$$x^{k+1}(t+1) = x_0 ,$$
$$x^{k+1}(t+1) = x^{k+1}(t) + f(x^{k+1}(t), u^{k+1}(t))$$
$$\text{for } t = 0, \ldots, T-1$$

and we proceed to the next step of the procedure. Concerning convergence of this method we can prove

Theorem 3.1. *If for every* $t \in \{0, \ldots, T-1\}$ *there is some*

$$u(t) \in U \text{ with } u(t) = \lim_{k \to \infty} u^k(t) ,$$

then $(u(0), \ldots, u(T-1)) \in U^T$ *is a solution of the problem (3.10).*

Proof. For $t = 0$ it follows from

$$x^{k+1}(1) = x_0 + f(x_0, u^{k+1}(0))$$

that the limit

$$\lim_{k \to \infty} x^{k+1}(1) = x(1) = x_0 + f(x_0, u(0))$$

exists. We assume that, for some $t \in \{0, \ldots, T-1\}$, the limit

$$\lim_{k \to \infty} x^{k+1}(t) = x(t) = x(t-1) + f(x(t-1), u(t-1))$$

exists. Then it follows from

$$x^{k+1}(t+1) = x^{k+1}(t) + f(x^{k+1}(t), u^{k+1}(t))$$

that the limit

$$\lim_{k \to \infty} x^{k+1}(t+1) = x(t+1) = x(t) + f(x(t), u(t))$$

exists.

By the principle of induction it therefore follows that the limit

$$\lim_{k \to \infty} x^{k+1}(t+1) = x(t+1)$$

exists and is given by

$$x(t+1) = x(t) + f(x(t), u(t))$$

for every $t \in \{0, \ldots, T-1\}$.
This implies because of

$$\tilde{x}^{k+1}(T) = x_0 + \sum_{t=0}^{T-1} f(x^k(t), u^{k+1}(t))$$

that

$$\lim_{k \to \infty} \tilde{x}^{k+1}(T) = x_0 + \sum_{t=0}^{T-1} f(x(t), u(t)) = x(T)$$

and hence

$$\lim_{k \to \infty} \varphi_k^T(u^{k+1}(0), \ldots, u^{k+1}(T-1)) = \varphi^T(u(0), \ldots, u(T-1)).$$

Now let $(\tilde{u}(0), \ldots, \tilde{u}(T-1))^T$ be chosen arbitrarily. Then it also follows that

$$\lim_{k \to \infty} \varphi_k^T(\tilde{u}(0), \ldots, \tilde{u}(T-1)) = \varphi^T(\tilde{u}(0), \ldots, \tilde{u}(T-1)) .$$

Further we have, for every $k \in \mathbb{N}_0$,

$$\varphi_k^T(\tilde{u}^{k+1}(0), \ldots, \tilde{u}^{k+1}(T-1)) \le \varphi_k^T(\tilde{u}(0), \ldots, \tilde{u}(T-1))$$

and hence

$$\varphi^T(u(0), \ldots, u(T-1)) = \lim_{k \to \infty} \varphi_k^T(u^{k+1}(0), \ldots, u^{k+1}(T-1)) \le$$
$$\lim_{k \to \infty} \varphi_k^T(\tilde{u}(0), \ldots, \tilde{u}(T-1)) = \varphi^T(\tilde{u}(0), \ldots, \tilde{u}(T-1))$$

This shows that $(u(0), \ldots, u(T-1)) \in U^T$ is a solution of the problem (3.10).

$\square$

We shall see later (see *Section 3.2.3*) that this procedure is not needed in the linear case where the functions $g_i : \mathbb{R}^N \times \mathbb{R}^M \to \mathbb{R}^{n_i}$ for $i = 1, \ldots, n$ in (3.1) are given by

$$g_i(x, u) = A_i x + B_i u , \quad x \in \mathbb{R}^N, \ u \in \mathbb{R}^M ,$$

with $n_i \times N$-matrices A_i and $n_i \times M$-matrices B_i.

3.2.2 The Non-Cooperative Case

For a given vector function $u : \mathbb{N}_0 \to \mathbb{R}^M$ we define, for every $i = 1, \ldots, n$ and $T \in \mathbb{N}$,

$$u_{T_i} = (u_i(0)^T, \ldots, u_i(T-1)^T)^T \in \mathbb{R}^{m_i \cdot T}$$

and

$$\tilde{G}_i^T (u_{T_1}, \ldots, u_{T_n}) = G_i^T (x_0, u(0), \ldots, u(T-1))$$

as well as

$$\tilde{\varphi}_i^T (u_{T_1}, \ldots, u_{T_n}) = \| \tilde{G}_i^T (u_{T_1}, \ldots, u_{T_n}) - \hat{x}_i \|_2^2 .$$

The system (3.9) can then be rewritten in the form

$$\tilde{G}_i^T (u_{T_1}, \ldots, u_{T_n}) = \hat{x}_i \text{ for } i = 1, \ldots, n . \tag{3.9'}$$

To solve system (3.9') non-cooperatively then means to find $u_{T_i}^* \in U_i^T$ for $i = 1, \ldots, n$ such that, for every $i = 1, \ldots, n$,

$$\tilde{\varphi}_i^T (u_{T_1}^*, \ldots, u_{T_n}^*) \leq \tilde{\varphi}_i^T (u_{T_1}^*, \ldots, u_{T_{i-1}}^*, u_{T_i}, u_{T_{i+1}}^*, u_{T_n}^*)$$

$$\text{for all } u_{T_i} \in U_i^T = \underbrace{U_i \times \ldots \times U_i}_{T-times} . \tag{3.13}$$

Every such n-tupel $(u_{T_1}^*, \ldots, u_{T_n}^*)$ is called a *Nash equilibrium*. In words (3.13) means that, if one player declines from his equilibrium control whereas all the others stick to it, his cost function cannot decrease.

Concerning the existence of a Nash equilibrium we can prove the

Theorem 3.2.
Assumptions:

a) *For every $i = 1, \ldots, n$ the control set U_i is convex and compact.*
b) *For every $i = 1, \ldots, n$ the vector function $g_i : \mathbb{R}^N \times \mathbb{R}^M \to \mathbb{R}^{n_i}$ in (3.1) is continuous.*
c) *For every n-tupel $(u_{T_1}^*, \ldots, u_{T_n}^*) \in U^T = \prod_{i=1}^{n} U_i^T$ and every $i = 1, \ldots, n$ there is exactly one*

$$S_i(\underbrace{u_{T_1}^*, \ldots, u_{T_n}^*}_{u_T^*}) = (u_{T_1}^*, \ldots, u_{T_{i-1}}^*, (S_i u_T^*)_i, u_{T_{i+1}}^*, \ldots, u_{T_n}^*)$$

with

$$\tilde{\varphi}_i^T (S_i(u_T^*)) \leq \tilde{\varphi}_i^T (u_{T_1}^*, \ldots, u_{T_{i-1}}^*, u_{T_i}, u_{T_{i+1}}^*, \ldots, u_{T_n}^*)$$

$$\text{for all } u_{T_i} \in U_i^T .$$

Assertion: There exists a Nash equilibrium.

Proof. From a), b), c), it follows that the mapping $S = S_n \circ S_{n-1} \circ \ldots \circ S_1 :$ $U^T \to U^T$ is continuous, since every mapping $S_i : U^T \to U^T$ is continuous which can be seen as follows:

Let $((u_{T_1}^k, \ldots, u_{T_n}^k))_{k \in \mathbb{N}}$ be a sequence in U^T with $u_T^k = (u_{T_1}^k, \ldots, u_{T_n}^k) \to (u_{T_1}^*, \ldots, u_{T_n}^*) \in U^T$.

Then, for every $k \in \mathbb{N}$ and every $i \in \{1, \ldots, n\}$ we have

$$\tilde{\varphi}_i(u_{T_1}^k, \ldots, u_{T_{i-1}}^k, (S_i u_T^k)_i, u_{T_{i+1}}^k, \ldots, u_{T_n}^k)$$
$$\leq \tilde{\varphi}_i^T(u_{T_1}^k, \ldots, u_{T_{i-1}}^k, u_{T_i}, u_{T_{i+1}}^k, \ldots, u_{T_n}^k) \qquad (3.14)$$
$$\text{for all } u_{T_i} \in U_i .$$

Further we have, for every $i = 1, \ldots, n$,

$$\tilde{\varphi}_i(u_{T_1}^*, \ldots, u_{T_{i-1}}^*, (S_i u_T^*)_i, u_{T_{i+1}}^*, \ldots, u_{T_n}^*)$$
$$\leq \tilde{\varphi}_i^T(u_{T_1}^*, \ldots, u_{T_{i-1}}^*, u_{T_i}, u_{T_{i+1}}^*, \ldots, u_{T_n}^*) \text{ for all } u_{T_i} \in U_i^T .$$

Now let $i \in \{1, \ldots, n\}$ be chosen arbitrarily. Then there is a subsequence $((u_{T_1}^{k_l}, \ldots, u_{T_n}^{k_l}))_{l \in \mathbb{N}}$ and a $\tilde{u}_{T_i} \in U_i^T$ with

$$\lim_{l \to \infty} (S_i u_T^{k_l})_i = \tilde{u}_{T_i} .$$

From (3.14) it follows therefore that

$$\tilde{\varphi}_i(u_{T_1}^*, \ldots, u_{T_{i-1}}^*, (S_i u_T^*)_i, u_{T_{i+1}}^*, \ldots, u_{T_n}^*)$$
$$\leq \tilde{\varphi}_i^T(u_{T_1}^*, \ldots, u_{T_{i-1}}^*, u_{T_i}, u_{T_{i+1}}^*, \ldots, u_{T_n}^*) \text{ for all } u_{T_i} \in U_i^T .$$

which implies (because of c)) that

$$\tilde{u}_{T_i} = (S_i u_T^*)_i .$$

In the same way one shows that, for every subsequence $((u_{T_1}^{k_l}, \ldots, u_{T_n}^{k_l}))_{l \in \mathbb{N}}$ there exists a subsequence $(u_{T_1}^{k_{l_m}}, \ldots, u_{T_n}^{k_{l_m}})_{m \in \mathbb{N}}$ such that

$$\lim_{m \to \infty} (S_i u_T^{k_{l_m}})_i = (S_i u_T^*)_i$$

which implies

$$\lim_{m \to \infty} (S_i u_T^k)_i = (S_i u_T^*)_i .$$

This shows that $S_i : U^T \to U^T$ is continuous for every $i \in \{1, \ldots, n\}$. From a) it follows that $U^T = \prod_{i=1}^n U_i^T$ is convex and compact.

By Brouwer's fixed point theorem $S = S_n \circ S_{n-1} \circ \ldots \circ S_1 : U^T \to U^T$ has a fixed point in U^T and every such is a Nash equilibrium.

This completes the proof of *Theorem 3.2*.

$\square$

In order to calculate a Nash equilibrium one can apply the following iteration method. Starting with $k = 0$ and $i = 1$ as well as some $u_T^k \in U^T$ a $u_{T_i}^* \in U_i^T$ is determined with

$$
\begin{aligned}
&\tilde{\varphi}_i^T (u_{T_1}^k, \ldots, u_{T_{i-1}}^k, u_{T_i}^*, u_{T_{i+1}}^k, \ldots, u_{T_n}^k) \\
&\leq \tilde{\varphi}_i^T (u_{T_1}^k, \ldots, u_{T_{i-1}}^k, u_{T_i}, u_{T_{i+1}}^k, \ldots, u_{T_n}^k) \text{ for all } u_{T_i} \in U_i^T .
\end{aligned}
\tag{3.15}
$$

Then one puts

$$
u_T^{k+1} = (u_{T_1}^k, \ldots, u_{T_{i-1}}^k, u_{T_i}^*, u_{T_{i+1}}^k, \ldots, u_{T_n}^k)
$$

and replaces i by $i + 1$ *modulo* n.
Concerning convergence we can prove the

Theorem 3.3.

Assumptions:

a) *For every $i = 1, \ldots, n$ the control set U_i^T is compact.*
b) *For every $i = 1, \ldots, n$ the vector function $g_i : \mathbb{R}^N \times \mathbb{R}^M \to \mathbb{R}^{n_i}$ in (3.1) is continuous.*
c) *The sequence $(u_T^k)_{k \in \mathbb{N}_0}$ converges to some $\hat{u}_T \in \mathbb{R}^{M \cdot T}$.*

Assertion: $\hat{u}_T$ is a Nash equilibrium.

Proof. From a) it follows that $U^T = \prod_{i=1}^n U_i^T$ is compact and hence closed which implies that $\hat{u}_T \in U^T$.
Let us assume that, for some $i \in \{1, \ldots, n\}$, there exists some $u_{T_i} \in U_i^T$ with

$$
\tilde{\varphi}^T (\hat{u}_{T_1}, \ldots, \hat{u}_{T_{i-1}}, u_{T_i}, \hat{u}_{T_{i+1}}, \ldots, \hat{u}_{T_n}) < \tilde{\varphi}_i^{\,T} (\hat{u}_T).
$$

Since the function $u \to \tilde{\varphi}_i^T (u)$, $u \in \mathbb{R}^{M \cdot T}$, is continuous, it follows that $\tilde{\varphi}_i^T (u_T^k) \to \tilde{\varphi}_i^T (\hat{u}_T)$. This implies for

$$
\delta = \frac{1}{2}(\tilde{\varphi}_i^T (\hat{u}_T) - \tilde{\varphi}_i^T (\hat{u}_{T_1}, \ldots, \hat{u}_{T_{i-1}}, u_{T_i}, \hat{u}_{T_{i+1}}, \ldots, \hat{u}_{T_n}))
$$

that

$$
\tilde{\varphi}_i^T (\hat{u}_T^{k+1}) > \tilde{\varphi}_i^T (\hat{u}_{T_1}, \ldots, \hat{u}_{T_{i-1}}, u_{T_i}, \hat{u}_{T_{i+1}}, \ldots, \hat{u}_{T_n}) + \delta
$$

$$
\text{for all } k \geq k_1(\delta) .
$$

If one puts, for every $k \in \mathbb{N}$,

$$
(u_T^k)^i = (u_{T_1}^k, \ldots, u_{T_{i-1}}^k, u_{T_i}, u_{T_{i+1}}^k, \ldots, u_{T_n}^k) ,
$$

then it follows that

$$\lim_{k \to \infty} (u_T^k)^i = (\hat{u}_{T_1}, \dots, \hat{u}_{T_{i-1}}, u_{T_i}, \hat{u}_{T_{i+1}}, \dots, \hat{u}_{T_n})$$

and hence

$$\tilde{\varphi}_i^T ((u_T^k)^i) < \tilde{\varphi}_i^T (\hat{u}_{T_1}, \dots, \hat{u}_{T_{i-1}}, u_{T_i}, \hat{u}_{T_{i+1}}, \dots, \hat{u}_{T_n}) + \delta$$

$$\text{for all } k \geq k_2(\delta) \geq k_1(\delta)$$

which contradicts (3.15).

$\square$

Therefore the assumption is false and $\hat{u}_T \in U^T$ is a Nash equilibrium. In order to solve the problem (3.15) for a given $i \in \{1, \dots, n\}$, $k \in \mathbb{N}$, and $u_T^k \in U^T$ we again apply the iterative procedure being described in *Section 3.2.1* under the assumption that the vector function $g : \mathbb{R}^N \times \mathbb{R}^M \to \mathbb{R}^N$ in (3.2) has the form

$$g(x, u) = x + f(x, u) , \quad x \in \mathbb{R}^N , \quad u \in \mathbb{R}^M .$$

For this purpose we consider the problem to find, for a given $i \in \{1, \dots, n\}$ and a given $u_T^* \in U^T$, some $\hat{u}_{T_i} \in U_i^T$ such that

$$\tilde{\varphi}_i^T (u_{T_1}^*, \dots, u_{T_{i-1}}^*, \hat{u}_{T_i}, u_{T_{i+1}}^*, \dots, u_{T_m}^*)$$
$$\leq \tilde{\varphi}_i^T (u_{T_1}^*, \dots, u_{T_{i-1}}^*, u_{T_i}, u_{T_{i+1}}^*, \dots, u_{T_m}^*)$$
$$\text{for all } u_{T_i} \in U_i^T .$$

At the beginning of the procedure we choose

$$u_i^0(t) = \Theta_{m_i} \text{ for all } t = 0, \dots, T - 1$$

and calculate

$$x^0(t+1) = x^0(t) + f(x^0(t), u_1^*(t), \dots, u_{i-1}^*(t), u_i^0(t), u_{i+1}^*(t), \dots, u_n^*(t))$$
$$\text{for } t = 0, \dots, T - 1$$

where

$$x^0(0) = x_0 .$$

Then we construct a sequence $(u_i^l(0), \dots, u_i^l(T-1))_{l \in \mathbb{N}_0}$ in U_i^T and a sequence $(x^l(1), \dots, x^l(T))_{l \in \mathbb{N}_0}$ in $\mathbb{R}^{N \cdot T}$ as follows:

If $(u_i^l(0), \ldots, u_i^l(T-1)) \in U_i^T$, $(x^l(1), \ldots, x^l(T)) \in \mathbb{R}^{N \cdot T}$ and $x^l(0) = x_0$ are given, then we determine $u_i^{l+1}(0), \ldots, u_i^{l+1}(T-1) \in U_i^T$ such that for

$$\tilde{x}^{l+1}(t+1) = \tilde{x}^l(t) + f(x^l(t), u_1^*(t), \ldots, u_{i-1}^*(t), u_i^{l+1}(t), u_{i+1}^*(t), \ldots, u_n^*(t))$$

for $t = 0, \ldots, T-1$ with $\tilde{x}^{l+1}(0) = x_0$ the function value

$$\tilde{\varphi}_{l,i}^T(u_{T_1}^*, \ldots, u_{T_{i-1}}^*, u_{T_i}^{l+1}, u_{T_{i+1}}^*, \ldots, u_{T_n}^*)$$

$$= \|\tilde{x}_i^{l+1} - \hat{x}_i\|_2^2$$

$$= \| \sum_{t=0}^{T-1} f_i(x^l(t), u_1^*(t), \ldots, u_{i-1}^*(t), u_i^{l+1}(t), u_{i+1}^*(t), \ldots, u_n^*(t)) \|_2^2$$

becomes minimal.
Then we define $x^{l+1}(0) = x_0$ and

$$x^{l+1}(t+1) =$$
$$x^{l+1}(t) + f(x^{l+1}(t), u_1^*(t), \ldots, u_{i-1}^*(t), u_i^{l+1}(t), u_{i+1}^*(t), \ldots, u_n^*(t))$$

for $t = 0, \ldots, T-1$ and proceed to the next step of the procedure.

3.2.3 The Linear Case

In this case the vector functions $g_i : \mathbb{R}^N \times \mathbb{R}^M \to \mathbb{R}^{n_i}$ for $i = 1, \ldots, n$ in (3.1) are given in the form

$$g_i(x, u) = A_i x + B_i u , \quad x \in \mathbb{R}^N , \quad u \in \mathbb{R}^M ,$$

with real $n_i \times N$-matrices A_i and $n_i \times M$-matrices B_i.
The system (3.2) then reads

$$x(t+1) = Ax(t) + Bu(t) , \quad t \in \mathbb{N}_0 , \tag{3.16}$$

where

$$A = \begin{pmatrix} A_1 \\ \vdots \\ A_n \end{pmatrix} \quad \text{and} \quad B = \begin{pmatrix} B_1 \\ \vdots \\ B_n \end{pmatrix} .$$

Obviously we have

$$g(\Theta_N, \Theta_M) = \Theta_N ,$$

i.e., $\hat{x} = \Theta_N$ is a fixed point of the uncontrolled system

$$x(t+1) = Ax(t) , \quad t \in \mathbb{N}_0 . \tag{3.17}$$

Further we obtain from (3.16) and

$$x(0) = x_0 \tag{3.18}$$

for some $x_0 \in \mathbb{R}^N$ that, for every $T \in \mathbb{N}$,

$$x(T) = A^T x_0 + \sum_{t=1}^{T} A^{T-t} Bu(t-1) ,$$

hence

$$G^T(x_0, u(0), \ldots, u(T-1)) = A^T x_0 + \sum_{t=1}^{T} A^{T-t} Bu(t-1)$$

so that the equation (3.9) reads

$$\sum_{t=1}^{T} A^{T-t} Bu(t-1) = -A^T x_0 . \tag{3.19}$$

Further we obtain as total cost function

$$\varphi^T(u(0), \ldots, u(T-1)) = \| \sum_{t=1}^{T} A^{T-t} Bu(t-1) + A^T x_0 \|_2^2$$

where $\| \cdot \|_2$ denotes the Euclidean norm in $\mathbb{R}^N$. Hence $\varphi^T : \mathbb{R}^{M \cdot T} \to \mathbb{R}$ is a convex functional and therefore automatically continuous.
If

$$rank(\, B \mid AB \mid \ldots \mid A^{T-1} B \,) = N \, ,$$

then φ^T is even strictly convex. If further the control sets U_i, $i = 1, \ldots, n$, are convex and compact, then the problem (3.10) has exactly one solution $(u^T(0)^T, \ldots, u^T(T-1)^T)^T \in U^T$ which, for instance, can be computed iteratively with the aid of the conditioned gradient method.
If the matrix A is non-singular, then the equation (3.19) can be replaced by

$$\sum_{t=1}^{T} A^{-t} Bu(t-1) = -x_0$$

and instead of φ^T the functional

$$\tilde{\varphi}^T(u(0), \ldots, u(T-1)) = \| \sum_{t=1}^{T} A^{-t} Bu(t-1) + x_0 \|_2^2$$

has to be minimized on U^T. In this case it follows that

$$\tilde{\varphi}^{T+1}(u^{T+1}(0), \ldots, u^{T+1}(T)) \le \tilde{\varphi}^{T+1}(u^T(0), \ldots, u^T(T-1), \Theta_M)$$

$$= \tilde{\varphi}^T(u^T(0), \ldots, u^T(T-1))$$

where again $(u^T(0), \ldots, u^T(T-1))$ denotes the solution of problem (3.10) with $\tilde{\varphi}^T$ instead of φ^T.
If one puts $C_t = A^{T-t} B$ for $t = 1, \ldots, T$ and defines

$$C_t = (\, C_{t1} \mid C_{t2} \mid \ldots \mid C_{tn} \,) \text{ with } N \times m_i \text{ matrices } C_{ti}$$
$$\text{for } i = 1, \ldots, n \, ,$$

then one obtains

$$\sum_{t=1}^{T} A^{T-t} Bu(t-1) = \sum_{t=1}^{T} C_t u(t-1) = \sum_{t=1}^{T} \sum_{i=1}^{n} C_{ti} u_i(t-1)$$

$$\sum_{i=1}^{n} \sum_{t=1}^{T} C_{ti} u_i(t-1) = \sum_{i=1}^{n} (\, C_{1i} \mid C_{2i} \mid \, \ldots \, \mid C_{Ti} \,) u_{Ti}$$

where $u_{Ti} = \left(u_i(0)^T, \ldots, u_i(T-1)^T \right)^T$ for $i = 1, \ldots, n$.

If we put $C(i) = (\, C_{1i} \mid C_{2i} \mid \, \ldots \, \mid C_{Ti} \,)$ for $i = 1, \ldots, n$, then it follows that

$$G^T(x_0, u(0), \ldots, u(T-1)) = A^T x_0 + \sum_{j=1}^{n} C(j) u_{Tj} \, ,$$

hence

$$\tilde{G}^T(u_{T1}, \ldots, u_{Tn}) = A^T x_0 + \sum_{j=1}^{n} C(j) u_{Tj} \, .$$

Let $C_i(j)$ be the i-th submatrix of $C(j)$ consisting of n_i row vectors of $C(j)$, i.e., let

$$C(j) = \begin{pmatrix} C_1(j) \\ \vdots \\ C_n(j) \end{pmatrix} \qquad \text{for } j = 1, \ldots, n$$

with $n_i \times (m_j \cdot T)$-matrices $C_i(j)$ and let

$$A^T x_0 = \begin{pmatrix} y_1 \\ \vdots \\ y_n \end{pmatrix} \qquad \text{with } y_i \in \mathbb{R}^{n_i} \text{ for } i = 1, \ldots, n \, .$$

Then it follows

$$\tilde{G}_i^T(u_{T1}, \ldots, u_{Tn}) = \sum_{j=1}^{n} C_i(j) u_{Tj} + y_i \text{ for } i = 1, \ldots, n$$

and the functional

$$\tilde{\varphi}_i^T(u_{T1}, \ldots, u_{Tn}) = \| \sum_{j=1}^{n} C_i(j) u_{Tj} + y_i \|_2^2$$

is convex (and hence continuous) on $\mathbb{R}^{M \cdot T}$. It is strictly convex, if

$$rank \; C_i(j) = n_i \text{ for all } j = 1, \ldots, n \, .$$

If this is the case for all $i = 1, \ldots, n$ and if the control sets U_i, $i = 1, \ldots, n$, are convex and compact, then all the assumptions of *Theorem 3.2* are satisfied and the existence of a Nash equilibrium is guaranteed.

3.3 Local Controllability

Let us come back to *Section 3.1* and assume that $g \in C^1(\mathbb{R}^N \times \mathbb{R}^M, \mathbb{R})$. Then we define

$$A = g_x(\hat{x}, \Theta_M) \quad \text{and} \quad B = g_u(\hat{x}, \Theta_M)$$

where $\hat{x}$ is a fixed point of the uncontrolled system (3.3).
Linearization of the system (3.2) at the point $(\hat{x}, \Theta_M)$ leads to the system

$$h(t + 1) = Ah(t) + Bu(t) , \quad t \in \mathbb{N}_0 , \qquad (3.20)$$

and the initial condition (3.18) has to be replaced by

$$h(0) = x_0 - \hat{x} . \qquad (3.21)$$

Instead of the problem of controllability we now consider the

Problem of Local Controllability

Given an initial state $x_0 \in \mathbb{R}^N$, find a time $T \in \mathbb{N}_0$ and a control function $u : \mathbb{N}_0 \to \mathbb{R}^M$ with

$$u(t) \in U = \prod_{i=1}^{n} U_i \quad \text{for all } t \in \mathbb{N}_0 \qquad (3.22)$$

and

$$u(t) = \Theta_M \quad \text{for all } t \geq T \qquad (3.23)$$

such that the solution $x : \mathbb{N}_0 \to \mathbb{R}^N$ of (3.20) which satisfies the initial condition (3.21) satisfies the end condition

$$h(t) = \Theta_N$$

which implies

$$h(t) = \Theta_N \quad \text{for all } t \geq T .$$

This problem is nothing else but the problem of null-controllability of the linear system (3.20) (see *Section 2.1.2*).
Let us assume, for every $i = 1, \ldots, n$, that

$$U_i = \{u \in \mathbb{R}^{n_i} \mid \|u\|_i \leq \gamma\}$$

where $\gamma > 0$ is a given constant and $\| \cdot \|_i$ is a norm in $\mathbb{R}^{n_i}$. Then we can prove (see *Theorem 2.2*) the following

Theorem 3.4. *Let there exist some $T \in \mathbb{N}$ such that*

$$rank(B \mid AB \mid \ldots \mid A^{T-1}B) = N .$$

Further let all the eigenvalues of $A' =$ transpose of A be less than or equal to one in absolute value and the corresponding eigenvectors be linearly independent. Then the problem of local controllability has a solution for every choice of $x_0 \in \mathbb{R}^N$, if A is non-singular.

3.4 An Emission Reduction Model

3.4.1 A Non-Cooperative Treatment

We come back to the emission reduction model that was introduced in *Section 1.1.6* as an uncontrolled system (see (1.22)) and in *Section 2.1.1* as a controlled system under an additional condition on the costs. Now we consider the actors who control the system as players who have to find a cost vector function $v : \mathbb{N}_0 \to \mathbb{R}^r$ with

$$\Theta_r \leq v(t) \leq M^* \text{ for } t = 0, \ldots, N-1 \, ,$$
$$v(t) = \Theta_r \text{ for } t \geq N \text{ for some } N \in \mathbb{N} \tag{3.24}$$

such that

$$C\left(\sum_{t=0}^{N-1} v(t)\right) = \hat{E} - E_0 \, , \ C = (em_{ij})_{i,j=1,\ldots,r} \ .$$

Let us replace this condition by

$$C\left(\sum_{t=0}^{N-1} v(t)\right) \geq \hat{E} - E_0 \tag{3.25}$$

and neglect the requirement

$$v(t) \leq M^* \text{ for } t = 0, \ldots, N-1$$

which in the case

$$M_i^* > 0 \text{ for } i = 1, \ldots, r$$

can always be satisfied, if we can find $v : \mathbb{N}_0 \to \mathbb{R}^r$ with (3.24), (3.25) (see *Section 2.1.1*). If we put

$$c_{ij} = em_{ij} \, , \ i, j = 1, \ldots, r \, , \ x = \sum_{t=0}^{N-1} v(t) \, , \ b = \hat{E} - E_0 \, ,$$

then (3.25) can be written in the form

$$\sum_{j=1}^{r} c_{ij} x_j \geq b_i, i = 1, \ldots, r \, , \tag{3.26}$$

and we have to find a vector $x \in \mathbb{R}^r$ with

$$x_i \geq 0 \text{ for } i = 1, \ldots, r \tag{3.27}$$

such that the inequalities (3.26) are satisfied.

Now each player is interested in minimizing his costs $x_i = \sum_{t=0}^{N-1} v_i(t)$, $i = 1,\ldots,r$. Let us assume that the players try to minimize the total costs

$$s(x) = \sum_{j=1}^{r} x_j \tag{3.28}$$

under the constraints (3.26), (3.27).

This is a typical problem of linear programming.

Let us assume that $\hat{x} \in \mathbb{R}^r$ is a solution of this problem. If we then choose, for any $i \in \{1,\ldots,r\}$, some $x_i \geq 0$ such that

$$\sum_{\substack{j=1 \\ j \neq i}}^{r} c_{kj}\hat{x}_j + c_{ki}x_i \geq b_k \text{ for } k = 1,\ldots,r ,$$

then it follows that

$$\sum_{j=1}^{r} \hat{x}_j \leq \sum_{\substack{j=1 \\ j \neq i}}^{r} \hat{x}_j + x_i ,$$

and therefore $\hat{x}_i \leq x_i$.

Thus every solution of (3.26), (3.27) which minimizes (3.28) is a Nash equilibrium, i.e., if the i-th player declines from his choice of costs whereas all the others stick to it, he can at most do worse.

The Dual Problem

The dual problem to the problem of minimizing (3.28) subject to (3.26), (3.27) consists of maximizing

$$t(y) = \sum_{i=1}^{r} b_i y_i , \quad y \in \mathbb{R}^r , \tag{3.29}$$

subject to

$$\sum_{i=1}^{r} c_{ij} y_i \leq 1 , \quad j = 1,\ldots,r , \tag{3.30}$$

and

$$y_i \geq 0 , \quad i = 1,\ldots,r . \tag{3.31}$$

For $y_1 = y_2 = \ldots = y_r = 0$ the side conditions (3.30), (3.31) are satisfied. If we assume that there exists some $x \in \mathbb{R}^r$ which satisfies (3.26), (3.27), then by a well known duality theorem there exists a solution $x = \hat{x} \in \mathbb{R}^r$ of (3.26), (3.27) which minimizes (3.28) and a solution $y = \hat{y} \in \mathbb{R}^r$ of (3.30), (3.31) which maximizes (3.29) and it is $s(\hat{x}) = t(\hat{y})$ which is equivalent to the two implications

$$
\text{(CSL)} \quad \left\{ \begin{array}{l} \\ \text{and} \\ \\ \end{array} \right.
\begin{array}{l}
\hat{x}_j > 0 \Rightarrow \displaystyle\sum_{i=1}^{r} c_{ij}\hat{y}_i = 1 \\
\\
\hat{y}_i > 0 \Rightarrow \displaystyle\sum_{i=1}^{r} c_{ij}\hat{x}_j = b_i \ .
\end{array}
$$

On introducing slack variables

$$z_j \geq 0 \quad \text{for } j = 1, \ldots, r \ , \tag{3.32}$$

condition (3.30) can be rewritten in the form

$$z_j + \sum_{i=1}^{r} c_{ij}y_i = 1 \ , \ j = 1, \ldots, r \ , \tag{3.33}$$

and the dual problem is equivalent to maximizing

$$\sum_{j=1}^{r} 0 \cdot z_j + \sum_{i=1}^{r} b_i y_i \tag{3.34}$$

subject to (3.31), (3.32), (3.33). This problem can be immediately solved with the aid of the simplex method starting with the feasible basis solution

$$z_j = 1 \ , \ j = 1, \ldots, r \ , \ \text{and} \ y_i = 0 \ , \ i = 1, \ldots, r \ .$$

Before proceeding we consider a
Special case
Let

$$b_j \geq 0 \quad \text{and} \quad c_{jj} > 0 \quad \text{for all } j = 1, \ldots, r \ .$$

If we assume that, for some $j \in \{1, \ldots, r\}$,

$$c_{ji} \leq 0 \quad \text{for all} \quad i = 1, \ldots, r \quad \text{with} \quad i \neq j \ ,$$

i.e., the player j can be considered as an opponent of all the others, then it follows for the solution $\hat{x} \in \mathbb{R}^r$ of (3.26), (3.27) which minimizes (3.28) that

$$\sum_{k=1}^{r} c_{jk}\hat{x}_k = b_j \ .$$

For otherwise $(\hat{x}_1, \ldots, \hat{x}_{j-1}, x_j^*, \hat{x}_{j+1}, \ldots, \hat{x}_r)$ with

$$x_j^* = \frac{1}{c_{jj}} \left(b_j - \sum_{\substack{k=1 \\ k \neq j}}^{r} c_{jk}\hat{x}_k \right) < \hat{x}_j$$

also solves (3.26), (3.27) and it follows $x_j^* + \sum\limits_{\substack{k=1 \\ k \neq j}}^{r} \hat{x}_k < \sum\limits_{k=1}^{r} \hat{x}_k$ contradicting

the minimality of $\sum\limits_{k=1}^{r} \hat{x}_k$. Now we assume that

$$c_{ji} \leq 0 \quad \text{for all } j \neq i \ ,$$

i.e., every player can be considered as an opponent of every other. Then it follows that

$$\sum\limits_{k=1}^{r} c_{jk} \hat{x}_k = b_j \quad \text{for all } j = 1, \ldots, r \ .$$

If in addition we assume that

$$\sum\limits_{j=1}^{r} c_{ij} > 0 \quad \text{for all } i = 1, \ldots, r \ ,$$

then

$$c_{jj} > 0 \quad \text{for all } j = 1, \ldots, r$$

and the matrix $C = (c_{ij})_{i,j=1,\ldots,r}$ is inverse monotone, i.e. C^{-1} exists and is positive (see L. Collatz: Funktionalanalysis und Numerische Mathematik. Springer-Verlag: Berlin, Göttingen, Heidelberg 1964). This implies

$$\hat{x} = C^{-1} b \geq \Theta_r \ .$$

If $x \in \mathbb{R}^r$ is any solution of (3.26), (3.27), then it follows that

$$x \geq C^{-1} b = \hat{x} \ , \quad \text{i.e. }, \quad x_i \geq \hat{x}_i \quad \text{for } i = 1, \ldots, r \ .$$

In words this means the following: If every player is an opponent of every other and if his own contribution to achieve his goal is greater than the negative sum of the contributions of his opponents, then everybody can reach the absolute minimum of his costs.

Now we return to the

General case

We assume that there exists a solution $x \in \mathbb{R}^r$ of (3.26), (3.27). Then the dual problem has a solution as seen above. If this has been obtained by $s \leq r$ steps of the simplex method, we can assume the result in the following form:

$$\begin{pmatrix} y_1 \\ \vdots \\ y_s \\ z_{s+1} \\ \vdots \\ z_r \end{pmatrix} = \begin{pmatrix} d_1 \\ \vdots \\ d_s \\ d_{s+1} \\ \vdots \\ d_r \end{pmatrix} + D \begin{pmatrix} -z_1 \\ \vdots \\ -z_s \\ -y_{s+1} \\ \vdots \\ -y_r \end{pmatrix}$$

where

$$D = \begin{pmatrix} d_{11} & \cdots & d_{1s} & d_{1s+1} & \cdots & d_{1r} \\ \vdots & & \vdots & \vdots & & \vdots \\ d_{s1} & \cdots & d_{ss} & d_{ss+1} & \cdots & d_{sr} \\ d_{s+11} & \cdots & d_{s+1s} & d_{s+1s+1} & \cdots & d_{s+1r} \\ \vdots & & \vdots & \vdots & & \vdots \\ d_{r1} & \cdots & d_{rs} & d_{rs+1} & \cdots & d_{rr} \end{pmatrix} ,$$

$$\sum_{j=1}^{r} b_j y_j = \sum_{j=1}^{s} b_j d_j + \sum_{k=1}^{s} \left(\sum_{j=1}^{s} d_{jk} b_j \right) (-z_k) + \sum_{k=s+1}^{r} \left(\sum_{j=1}^{s} d_{jk} b_j \right) (-y_k)$$

with

$$d_j \geq 0 \quad \text{for} \quad j = 1, \ldots, r .$$

and

$$\sum_{j=1}^{s} d_{jk} b_j \geq 0 \quad \text{for} \quad k = 1, \ldots, r .$$

The corresponding solution of the dual problem is given by

$$\hat{y}_j = d_j \quad \text{for } j = 1, \ldots, s \quad \text{and} \quad \hat{y}_j = 0 \quad \text{for } j = s+1, \ldots, r .$$

Further we have

$$d_j + \sum_{i=1}^{r} c_{ij} \hat{y}_i = 1 \quad \text{for } j = s+1, \ldots, r .$$

Let us assume that

$$d_j > 0 \quad \text{for all } j = 1, \ldots, s .$$

If $\hat{x} \in \mathbb{R}^r$ is any solution of (3.26), (3.27) which minimizes (3.28), then it follows from the implications (CSL) that

$$\hat{x}_j = 0 \quad \text{for } j = s+1, \ldots, r$$

and

$$\sum_{j=1}^{s} c_{ij} \hat{x}_j = b_i \quad \text{for } i = 1, \ldots, s .$$

If the matrix

$$C_s = \begin{pmatrix} c_{11} & \cdots & c_{1s} \\ \vdots & \ddots & \vdots \\ c_{s1} & \cdots & c_{ss} \end{pmatrix}$$

is invertible, it follows that

$$C_s^{-1} = \begin{pmatrix} d_{11} & \cdots & d_{s1} \\ \vdots & \ddots & \vdots \\ d_{1s} & \cdots & d_{ss} \end{pmatrix}$$

which implies for $\hat{x}^s = (\hat{x}_1, \ldots, \hat{x}_s)^T$ that

$$\hat{x}^s = \begin{pmatrix} d_{11} & \cdots & d_{s1} \\ \vdots & \ddots & \vdots \\ d_{1s} & \cdots & d_{ss} \end{pmatrix} \begin{pmatrix} b_1 \\ \vdots \\ b_s \end{pmatrix} ,$$

hence

$$\hat{x}_k = \sum_{j=1}^{s} d_{jk} b_j \quad \text{for } k = 1, \ldots, s .$$

Let us continue with a direct method for the determination of a Nash equilibrium, i.e., of an $\hat{x} \in \mathbb{R}^r$ with $\hat{x} \geq \Theta_r$ and

$$\sum_{j=1}^{r} c_{ij} \hat{x}_j \geq b_i , \quad i = 1, \ldots, r \tag{3.35}$$

such that the following is true:
If for an arbitrary $i \in \{1, \ldots, r\}$ there exists some $x_i \geq 0$ with

$$\sum_{\substack{j=1 \\ j \neq i}}^{r} c_{kj} \hat{x}_j + c_{ki} x_i \geq b_k , \quad k = 1, \ldots, r ,$$

then it follows that $\hat{x}_i \leq x_i$. In order to determine such a Nash equilibrium we apply an iterative method as follows:
Starting with a vector $x^0 \geq \Theta_r$ which satisfies (3.35) with x^0 instead of $\hat{x}$ we construct a sequence $(x^L)_{L \in \mathbb{N}_0}$ with $L = l \cdot r + i$, $l \in \mathbb{N}_0$, $i = 1, \ldots, n-1$ in the following manner: If $x^L \geq \Theta_r$ with (3.35) for x^L instead of $\hat{x}$ is given, then we minimize $x_i \in \mathbb{R}$ subject to $x_i \geq 0$ and

$$\sum_{\substack{j=1 \\ j \neq i}}^{r} c_{kj} x_j^L + c_{ki} x_i \geq b_k , \quad k = 1, \ldots, r . \tag{3.36}$$

This problem has a solution $x_i^* \geq 0$ which can be explicitly calculated if $c_{ii} > 0$ for all $i = 1, \ldots, r$, as we shall see later and for which $x_i^* \leq x_i^L$ holds true. If we define

$$x_j^{L+1} = \begin{cases} x_j^L & \text{for } j \neq i , \\ x_j^* & \text{for } j = i \end{cases}$$

where

$$L + 1 = \begin{cases} (l+1)r \, , & \text{if} \quad i = r - 1 \, , \\ l \cdot r + i + 1 \, , & \text{if} \quad i < r - 1 \, , \end{cases}$$

then $x^{L+1} \geq \Theta_r$ satisfies (3.35) with x^{L+1} instead of $\hat{x}$ and $x^{L+1} \leq x^L$. The latter implies the existence of

$$\hat{x} = \lim_{L \to \infty} x^L \leq x^L \quad \text{for all } L \in \mathbb{N}_0$$

which satisfies (3.35).

Assertion: $\hat{x}$ is a Nash equilibrium, if

$$c_{ii} > 0 \text{ for all } i \in \{1, \dots, r\} \tag{3.37}$$

Proof. Assume that $\hat{x}$ is not a Nash equilibrium. Then there is some $i \in \{1, \dots, r\}$ and some $x_i \geq 0$ such that

$$\sum_{\substack{j=1 \\ j \neq i}}^{r} c_{kj} \hat{x}_j + c_{ki} x_i \geq b_k \quad k = 1, \dots, \text{r}$$

and $x_i < \hat{x}_i$.
This implies

$$\sum_{\substack{j=1 \\ j \neq i}}^{r} c_{ij} \hat{x}_j + c_{ii} \hat{x}_i > b_i \tag{3.38}$$

If we define a subsequence $(L_l)_{l \in \mathbb{N}_0}$ by $L_l = l \cdot r + i$, then we obtain

$$\sum_{\substack{j=1 \\ j \neq i}}^{r} c_{kj} x_j^{L_l} + c_{ki} x_i^{L_l+1} \geq b_k$$

$$\text{for all } k = 1, \dots, r \quad \text{and all} \quad l \in \mathbb{N}_0 \, .$$

In particular it follows that

$$\sum_{\substack{j=1 \\ j \neq i}}^{r} c_{ij} x_j^{L_l} + c_{ii} x_i^{L_l+1} = b_i \quad \text{for all } l \in \mathbb{N}_0$$

(for otherwise $x_i^{L_l+1}$ could be chosen smaller) which implies

$$\sum_{\substack{j=1 \\ j \neq i}}^{r} c_{ij} \hat{x}_j + c_{ii} \hat{x}_i = b_i$$

contradicting (3.38). Hence the assumption is false and $\hat{x}$ is a Nash equilibrium.

$\square$

In order to minimize $x_i \geq 0$ subject to (3.36) we proceed as follows:

1. If $x_i^L = 0$, then we put $x_i^* = 0$ and are done.
2. If $x_i^L > 0$ and

$$\sum_{\substack{j=1 \\ j \neq i}}^{r} c_{ij} x_j^L + c_{ii} x_i^L = b_i \; ,$$

   we put $x_i^* = x_i^L$ and are done.
3. If

$$\sum_{\substack{j=1 \\ j \neq i}}^{r} c_{ij} x_j^L + c_{ii} x_i^L > b_i$$

   and there is some $k \neq i$ such that

$$c_{ki} > 0 \quad \text{and} \quad \sum_{\substack{j=1 \\ j \neq i}}^{r} c_{kj} x_j^L + c_{ki} x_i^L = b_k \; ,$$

   then we also put $x_i^* = x_i^L$ and are done.
4. Otherwise we have

$$c_{ki} \leq 0 \quad \text{for all} \quad k \in I(x^L)$$

   where

$$I(x^L) = \{ k \mid \sum_{\substack{j=1 \\ j \neq i}}^{r} c_{kj} x_j^L + c_{ki} x_i^L = b_k \} \; .$$

Let $J(L)$ be the complement of $I(x^L)$, i.e.,

$$J(L) = \{ k \mid \sum_{\substack{j=1 \\ j \neq i}}^{r} c_{kj} x_j^L + c_{ki} x_i^L > b_k \} \; .$$

Now let $h_i \leq x_i^L$ be such that

$$\sum_{\substack{j=1 \\ j \neq i}}^{r} c_{kj} x_j^L + c_{ki} (x_i^L - h_i) \geq b_k \quad \text{for all } k = 1, \ldots, r \; .$$

Then

$$h_i \leq \frac{1}{c_{ki}} \left(\sum_{\substack{j=1 \\ j \neq i}}^{r} c_{kj} x_j^L + c_{ki} x_i^L - b_k \right) = \alpha_k^L$$

for all $k \in J(L)$ with $c_{ki} > 0$.

If we therefore put $x_i^* = x_i^L - h_i$ with

$$h_i = \min \left(x_i^L, \min\{\alpha_k^L \mid k \in J(L) \quad \text{and} \quad c_{ki} > 0\} \right) ,$$

then $0 \le x_i^* \le x_i^L$ and x_i^* is the smallest non-negative number that satisfies (3.36).

In particular, if $c_{kj} \le 0$ for all $j \neq k$, then we get in the case

$$\sum_{\substack{j = 1 \\ j \neq i}}^{r} c_{ij}x_j^L + c_{ii}x_i^L > b_i$$

that

$$h_i = \min(x_i^L, \alpha_i^L)$$

where

$$\alpha_i^L = \frac{1}{c_{ii}} \left(\sum_{\substack{j = 1 \\ j \neq i}}^{r} c_{ij}x_j^L + c_{ii}x_i^L - b_i \right)$$

which implies

$$\sum_{\substack{j = 1 \\ j \neq i}}^{r} c_{ij}x_j^L + c_{ii}x_i^* = b_i .$$

Let us demonstrate this procedure by a numerical example: We choose

$$C = \begin{pmatrix} 1.667 & -0.875 & -0.792 \\ -0.792 & 1.667 & -0.875 \\ -0.167 & -0.167 & 0.333 \end{pmatrix} \times 10^{-2}$$

and

$$b = \begin{pmatrix} -0.3459 \\ -0.1083 \\ 0.0498 \end{pmatrix}$$

Starting with $x^0 = (0, 2, 16)^T$ which satisfies $Cx^0 \ge b$ we obtain the sequence

$$
\begin{array}{llll}
x_1^1 = 0 \,, & x_2^1 = & 2 & \,, x_3^1 = & 16 & ; \\
x_1^2 = 0 \,, & x_2^2 = 1.9016197 \,, & x_3^2 = & 16 & ; \\
x_1^3 = 0 \,, & x_2^3 = 1.9016197 \,, & x_3^3 = & 15.90862 & ; \\
x_1^4 = 0 \,, & x_2^4 = 1.9016197 \,, & x_3^4 = & 15.90862 & ; \\
x_1^5 = 0 \,, & x_2^5 = 1.8536548 \,, & x_3^5 = & 15.90862 & ; \\
x_1^6 = 0 \,, & x_2^6 = 1.8536548 \,, & x_3^6 = & 15.884566 & ; \\
x_1^7 = 0 \,, & x_2^7 = 1.8536548 \,, & x_3^7 = & 15.884566 & ; \\
x_1^8 = 0 \,, & x_2^8 = 1.8410289 \,, & x_3^8 = & 15.884566 & ; \\
x_1^9 = 0 \,, & x_2^9 = 1.8410289 \,, & x_3^9 = & 15.878234 & .
\end{array}
$$

The sequences $(x_2^L)_{L \in \mathbb{N}_0}$ and $(x_3^L)_{L \in \mathbb{N}_0}$ converge to the solutions x_2 and x_3 of the linear system

$$0.01667x_2 - 0.00875x_3 = -0.1083$$
$$-0.00167x_2 - 0.00333x_3 = 0.0498$$

which are approximately given by

$$x_2 = 1.8365109 \quad \text{and} \quad x_3 = 15.875959 \ .$$

3.4.2 A Cooperative Treatment

Let us assume that we have found a cost vector function $v : \mathbb{N}_0 \to \mathbb{R}^r$ which satisfies (3.24), (3.25). Then the controlled costs are given by

$$M_i(t+1) = v_i(t) - \lambda_i v_i(t)(M_i^* - v_i(t))E_i(t)$$

$$= v_i(t) - \lambda_i v_i(t)(M_i^* - v_i(t))(E_i(t-1) + \sum_{j=1}^{r} em_{ij}v_j(t-1))$$

for $i = 1, \ldots, r$ and $t \in \mathbb{N}$. Now let K be any subset of $\mathcal{N} = \{1, \ldots, r\}$ and, for any $t \in \{1, \ldots, N-1\}$, let $c^K(t-1) = (c_{ij}^K(t-1))_{i,j=1,\ldots,r}$ be a non-negative $r \times r$-matrix with

$$c_{ii}^K(t-1) = 0 \text{ for } i = 1, \ldots, r \qquad \text{and}$$
$$c_{ij}^K(t-1) > 0 \text{ for } i, j \in K(i \neq j) \ .$$

If we define, for every $t \in \{1, \ldots, N-1\}$,

$$\tilde{c}_{ij}^K(t-1) = em_{ij} + c_{ij}^K(t-1)$$

and

$$\tilde{c}_{ij}^K(N-1) = em_{ij} \text{ for } i, j = 1, \ldots, r \ ,$$

then

$$\sum_{t=1}^{N} \tilde{C}^K(t-1)v(t-1) \geq C(\sum_{t=0}^{N-1} v(t)) \geq \hat{E} - E_0 \ .$$

Hence the condition (3.25) is also satisfied, if we replace, for every $t \in \{0, \ldots, N-1\}$, the matrix C by $\tilde{C}^K(t) = (\tilde{c}_{ij}^K(t))_{i,j=1,\ldots,r}$. The controlled costs are then given by

$$M_i^K(t+1) = v_i(t) - \lambda_i v_i(t)(M_i^* - v_i(t))(E_i(t-1) + \sum_{j=1}^{r} \tilde{c}_{ij}^K(t-1)v_j(t-1))$$

for $i = 1, \ldots, r$ and $t = 1, \ldots, N-1$, and it follows that

$$M_i^K(t+1) \leq M_i(t+1) \text{ for all } i = 1, \ldots, r \text{ and } t = 1, \ldots, N-1 \ .$$

If we define, for every $K \subseteq \mathcal{N}$ and every $t \in \{1, \ldots, N-1\}$,
$v_t(K) =$

$$\sum_{i=1}^{r}(M_i(t+1) - M_i^K(t+1)) = \sum_{i=1}^{r}\lambda_i v_i(t)(M_i^* - v_i(t))\sum_{j=1}^{r}c_{ij}^K(t-1)v_j(t-1) ,$$

then

$$v_t(\phi) = 0 .$$

The function $v_t : 2^N \to \mathbb{R}_+$ can therefore be interpreted as the payoff function of a cooperative r-person game. The subsets K of $\mathcal{N}$ can be interpreted as coalitions which are built by the players by changing the matrix C of mutual influence to the matrix $\tilde{C}^K(t-1)$ for $t = 1, \ldots, N-1$ whereby they guarantee that the controlled costs are diminished. If $i \notin K$, then $c_{ij}^K(t-1) = 0$ for all $j \in \mathcal{N}$ and therefore $M_i(t+1) = M_i^K(t+1)$ so that

$$v_t(K) = \sum_{i \in K}(M_i(t+1) - M_i^K(t+1)) .$$

In particular we have

$$v_t(\mathcal{N}) = \sum_{i=1}^{r}(M_i(t+1) - M_i^{\mathcal{N}}(t+1)) .$$

If we denote the gain of the i-th player, if he joins the coalition $K \subseteq \mathcal{N}$, by

$$v_t^i(K) = M_i(t+1) - M_i^K(t+1) = \lambda_i v_i(t)(M_i^* - v_i(t))\sum_{j=1}^{r}c_{ij}^K(t-1)v_j(t-1)$$

then

$$v_t(K) = \sum_{i \in K}v_t^i(K) .$$

Let us assume that

$$v_t(\mathcal{N}) \geq v_t(K) \text{ for all } K \subseteq \mathcal{N} .$$

Then the "grand coalition" $\mathcal{N}$ leads to the largest joint gain $v_t(\mathcal{N})$. The question now is whether there exists a division $(x_1, \ldots, x_r)$ of $v_t(\mathcal{N})$, i.e., $x_i \geq 0$ for all $i = 1, \ldots, r$ and $v_t(\mathcal{N}) = \sum_{i=1}^{r}x_i$ such that

$$\sum_{i \in K}x_i \geq v_t(K) \text{ for all } K \subseteq \mathcal{N} .$$

This means that there is no incentive to build coalitions which differ from the grand coalition. The set of all such divisions of $v_t(\mathcal{N})$ is called the core of the game.

3.4.3 Conditions for the Core to be Non-Empty

The existence of a non-empty core is guaranteed, if

$$\sum_{j=1}^{r} c_{ij}^{K}(t-1)v_j(t-1) \le \sum_{j=1}^{r} c_{ij}^{\mathcal{N}}(t-1)v_j(t-1)$$

for all $i = 1, \ldots, r$ and $K \subseteq \mathcal{N}$.

From this condition it namely follows that

$$v_t^i(K) \le v_t^i(\mathcal{N}) \text{ for } i = 1, \ldots, r \text{ and all } K \subseteq \mathcal{N} .$$

If we therefore put

$$x_i = v_t^i(\mathcal{N}) \text{ for } i = 1, \ldots, r ,$$

then we can conclude that

$$x_i \ge 0 \text{ for } i = 1, \ldots, r ,$$

$$\sum_{i=1}^{r} x_i = v_t(\mathcal{N}) \text{ and } \sum_{i \in K} x_i \ge v_t(K) \text{ for all } K \subseteq \mathcal{N} ,$$

i.e. $(x_1, \ldots, x_r)^T$ is in the core of v_t.
In general it is not easy to show that the core $c(v_t)$ of v_t is not empty. In order to get some more insight of its structure we give another definition of the core, however, for so called superadditive games which have the property that

$$v_t(K \cup L) \ge v_t(K) + v_t(L) \text{ for all } K, L \subseteq \mathcal{N} \text{ with } K \cap L = \phi .$$

For this purpose we define the set of all divisions of $v_t(\mathcal{N})$ by $I(v_t)$, i.e.,

$$I(v_t) = \{x \in \mathbb{R}^r \mid x_i \ge 0 \text{ for } i = 1, \ldots, r \text{ and } \sum_{i=1}^{r} x_i = v_t(\mathcal{N})\} .$$

We say that $x \in I(v_t)$ dominates $y \in I(v_t)$, if there exists a coalition $K \subseteq \mathcal{N}$ such that

$$x_i > y_i \text{ for all } i \in K$$

$$(*)$$

$$\text{and } \sum_{i \in K} x_i \le v_t(K) .$$

Then we can prove that

$$c(v_t) = \{y \in I(v_t) \mid \textit{There is no } x \in I(v_t) \textit{ that dominates } y\} .$$

Proof. First of all we observe that $c(v_t) \subseteq I(v_t)$. Now let $y \in c(v_t)$. Assume that there exists some $x \in I(v_t)$ that dominates y. Then there is a coalition $K \subseteq \mathcal{N}$ such that (*) holds true which implies

$$\sum_{i \in K} y_i < \sum_{i \in K} x_i \leq v_t(K)$$

and contradicts the assumption that $y \in c(v_t)$.

Now let $y \in I(v_t) \setminus c(v_t)$. Then there exists a coalition $K \subseteq \mathcal{N}$ with $K \neq \emptyset$, $\mathcal{N}$ and $\sum\limits_{i \in K} y_i < v_t(K)$. Then we define

$$\rho = v_t(\mathcal{N}) - v_t(K) - \sum_{i \in \mathcal{N} \setminus K} v_t^i(\mathcal{N} \setminus K) \,,$$

$$\sigma = v_t(K) - \sum_{i \in K} y_i$$

and put

$$x_i = \begin{cases} y_i + \frac{\sigma}{|K|} & \text{for } i \in K \,, \\ v_t^i(\mathcal{N} \setminus K) + \frac{\rho}{|\mathcal{N} \setminus K|} & \text{for } i \in \mathcal{N} \setminus K \,. \end{cases}$$

Because of $\sigma > 0$ it follows that $x_i > y_i$ for all $i \in K$. Further it follows that

$$\sum_{i \in K} x_i = \sum_{i \in K} y_i + v_t(K) - \sum_{i \in K} y_i = v_t(K)$$

and

$$\sum_{i=1}^r x_i = v_t(K) + \sum_{i \in \mathcal{N} \setminus K} v_t^i(\mathcal{N} \setminus K) + v_t(\mathcal{N}) - v_t(K) - \sum_{i \in \mathcal{N} \setminus K} v_t^i(\mathcal{N} \setminus K) = v_t(\mathcal{N}) \,.$$

Since

$$v_t(\mathcal{N}) \geq v_t(K) + v_t(\mathcal{N} \setminus K) = v_t(K) + \sum_{i \in \mathcal{N} \setminus K} v_t^i(\mathcal{N} \setminus K) \,,$$

it follows that $\rho \geq 0$ and

$$x_i \geq v_t^i(\mathcal{N} \setminus K) \geq 0 \quad \text{for } i \in \mathcal{N} \setminus K \,.$$

Therefore $x = (x_1, \dots, x_r)^T \in I(v_t)$ dominates y which completes the proof.

$\square$

Next we give a necessary condition for the core to be non-empty. For that purpose we define

$$b_i(v_t) = v_t(\mathcal{N}) - v_t(\mathcal{N} \setminus \{i\}) \quad \text{for } i = 1, \dots, r$$

and see that

$$b_i(v_t) \geq 0 \quad \text{for all } i = 1, \ldots, r \ .$$

Now let $x \in c(v_t)$. Then it follows that

$$b_i(v_t) = \sum_{j=1}^{r} x_j - v_t(\mathcal{N}\backslash\{i\}) \geq \sum_{j=1}^{r} x_j - \sum_{\substack{j=1 \\ j \neq i}}^{r} x_j = x_i$$

$$\text{for all } i = 1, \ldots, r \ .$$

Let us define a gap function by

$$g_t(S) = \sum_{j \in S} b_j(v_t) - v_t(S) \text{ for } S \subseteq \mathcal{N} \ .$$

Then it follows, for every $S \subseteq \mathcal{N}$, that

$$g_t(S) \geq \sum_{j \in S} x_j - v_t(S) \geq 0 \ .$$

This is therefore a necessary condition for the core to be non-empty. In the following we assume this condition to be satisfied. If we define, for every $i = 1, \ldots, r$,

$$\lambda_i(v_t) = min\{g_t(S) \mid S \subseteq \mathcal{N} \text{ with } i \in S\} \ ,$$

it follows that

$$b_i(v_t) \geq \lambda_i(v_t) \geq 0 \text{ for all } i = 1, \ldots, r \ .$$

Further we obtain, for every $x \in c(v_t)$,

$$\lambda_i(v_t) = g_t(S^*) \text{ for some } S^* \subseteq \mathcal{N} \text{ with } i \in S^*$$
$$= \sum_{j \in S^*} b_j(v_t) - v_t(S^*)$$
$$\geq \sum_{j \in S^*} (b_j(v_t) - x_j) \geq b_i(v_t) - x_i \ ,$$

hence

$$b_i(v_t) - \lambda_i(v_t) \leq x_i \leq b_i(v_t)$$
$$\text{for all } i = \{1, \ldots, r\} \ .$$

If in addition to

$$g_t(S) = \sum_{j \in S} b_j(v_t) - v_t(S) \geq 0 \text{ for all } S \subseteq \mathcal{N}$$

we assume that

$$g_t(\mathcal{N}) = \sum_{j=1}^{r} b_j(v_t) - v_t(\mathcal{N}) = 0 \ ,$$

then it follows that $(b_1(v_t), \ldots, b_r(v_t))^T \in c(v_t)$.
So these two conditions are sufficient for the core to be non-empty.
If $g_t(\mathcal{N}) > 0$ and $\sum_{j=1}^{r} \lambda_i(v_t) > 0$, then we define

$$\tau_i(v_t) = b_i(v_t) - \frac{g_t(\mathcal{N})}{\sum_{j=1}^{r} \lambda_j(v_t)} \lambda_i(v_t) \text{ for } i = 1, \ldots, r$$

and obtain

$$\tau_i(v_t) \geq 0 \text{ for } i = 1, \ldots, r \ ,$$

if

$$\sum_{j=1}^{r} \lambda_j(v_t) \geq g_t(\mathcal{N}) \ .$$

Further we obtain

$$\sum_{i=1}^{r} \tau_i(v_t) = \sum_{i=1}^{r} b_i(v_t) - \sum_{j=1}^{r} b_j(v_t) + v_t(\mathcal{N}) = v_t(\mathcal{N})$$

and

$$\sum_{i \in S} \tau_i(v_t) = \sum_{i \in S} (b_i(v_t) - \frac{g_t(\mathcal{N})}{\sum_{j=1}^{r} \lambda_j(v_t)} \lambda_i(v_t)) \geq v_t(S) \ ,$$

if

$$g_t(S) - \frac{g_t(\mathcal{N})}{\sum_{j=1}^{r} \lambda_j(v_t)} \sum_{i \in S} \lambda_i(v_t) \geq 0 \text{ for all } S \subseteq \mathcal{N}$$

which implies that $(\tau_1(v_t), \ldots, \tau_r(v_t))^T \in c(v_t)$.
Assume that

$$\lambda_i(v_t) = g_t(\mathcal{N}) \text{ for all } i = 1, \ldots, r \ .$$

Then the last condition reads

$$g_t(S) - \frac{|S|}{r} g_t(\mathcal{N}) \geq 0 \text{ for all } S \subseteq \mathcal{N}$$

and we obtain

$$\tau_i(v_t) = b_i(v_t) - \frac{1}{r} g_t(\mathcal{N}) \text{ for } i = 1, \ldots, r .$$

Further

$$\sum_{j=1}^{r} \lambda_j(v_t) = r \cdot g_t(\mathcal{N}) \geq g_t(\mathcal{N}) > 0$$

is satisfied.

Result. If

$$\lambda_i(v_t) = g_t(\mathcal{N}) > 0 \text{ for all } i = 1, \ldots, r$$

and

$$g_t(S) - \frac{|S|}{r} g_t(\mathcal{N}) \geq 0 \text{ for all } S \subseteq \mathcal{N} ,$$

then

$$\left(b_1(v_t) - \frac{1}{r} g_t(\mathcal{N}), \ldots, b_r(v_t) - \frac{1}{r} g_t(\mathcal{N})\right) \in c(v_t) .$$

Remark: If $g_t(\mathcal{N}) = 0$, then this result coincides with the one obtained above.

3.4.4 Further Conditions for the Core to be Non-Empty

In the following we replace r by n. Next we will present a constructive method by which we can decide whether the core $c(v_t)$ of v_t is empty or not. For this purpose we order the subsets of $\mathcal{N}$ which have at least two elements in a sequence $K_1, \ldots, K_{2^n - n - 1}$ such that

$$|K_i| \leq |K_{i+1}| \text{ for } i = 1, \ldots, 2^n - n - 2$$

which implies that $K_{2^n - n - 1} = \mathcal{N}$.

Then we define a $(2^n - n - 1) \times n$-matrix $A = (a_{ik})_{\substack{i = 1, \ldots, 2^n - n - 1 \\ k = 1, \ldots, n}}$ by

$$a_{ik} = \begin{cases} 0 & \text{if } k \notin K_i , \\ 1 & \text{if } k \in K_i . \end{cases}$$

and a $(2^n - n - 1)$-vector $b = (b_i)_{i=1,\ldots,2^n - n - 1}$ by

$$b_i = v_t(K_i) \text{ for } i = 1, \ldots, 2^n - n - 1 .$$

With these definitions we conclude that a vector $x \in \mathbb{R}^n$ is in the core $c(v_t)$, if and only if

$$\sum_{k=1}^{n} a_{ik} x_k \geq b_i \text{ for } i = 1, \dots, 2^n - n - 2, \quad (P1)$$

$$\sum_{k=1}^{n} x_k = b_{2^n - n - 1} = v_t(\mathcal{N}) \quad (P2)$$

and

$$x_k \geq 0 \text{ for } k = 1, \dots, n . \quad (P3)$$

Now we replace the constraint $(P1)$ by

$$\sum_{k=1}^{n} a_{ik} x_k + x_{n+1} \geq b_i \text{ for } i = 1, \dots, 2^n - n - 2 , \quad (P1')$$

and consider the problem of minimizing x_{n+1} subject to $(P1')$, $(P2)$ and

$$x_k \geq 0 \text{ for } k = 1, \dots, n + 1 . \quad (P3')$$

This is a problem of linear programming whose dual problem consists of maximizing

$$\sum_{i=1}^{2^n - n - 1} b_i y_i$$

subject to the constraints

$$\sum_{i=1}^{2^n - n - 1} a_{ik} y_i \leq 0 \qquad \text{for } k = 1, \dots, n , \quad (D1)$$

$$\sum_{i=1}^{2^n - n - 2} y_i \leq 1 , \quad (D2)$$

$$y_i \geq 0 \qquad \text{for } i = 1, \dots, 2^n - n - 2 . \quad (D3)$$

If we choose $(x_1, \dots, x_k)$ such that $(P2)$ and $(P3)$ are satisfied (which is possible), then we can choose $x_{n+1} \geq 0$ large enough so that $(P1')$ is satisfied. Thus we can find $x_1, \dots, x_{n+1}$ such that the constraints $(P1')$, $(P2)$ and $(P3')$ are satisfied. If we choose $y_i = 0$ for $i = 1, \dots, 2^n - n - 1$, then the constraints $(D1)$, $(D2)$, $(D3)$ are also satisfied. By a well known duality theorem there are numbers $\hat{x}_1, \dots, \hat{x}_{n+1}$ which satisfy $(P1')$, $(P2)$, $(P3)$ and are such that $\hat{x}_{n+1} \geq 0$ is minimal and there are numbers $\hat{y}_i$ for $i = 1, \dots, 2^n - n - 1$ such that $(D1)$, $(D2)$, $(D3)$ hold true and $\sum_{i=1}^{2^n - n - 1} b_i \hat{y}_i$ is maximal. Moreover,

$$\sum_{i=1}^{2^n - n - 1} b_i \hat{y}_i = \hat{x}_{n+1} \geq 0 .$$

This implies that the core $c(v_t)$ of v_t is non-empty, if and only if $\hat{x}_{n+1} = 0$ or if and only if for every set of numbers y_i, $i = 1, \ldots, 2^n - n - 1$, with $(D1)$ and $(D3)$ it follows that

$$\sum_{i=1}^{2^n - n - 1} b_i y_i \leq 0 .$$

Let us consider the case $n = 3$. Then the constraints $(P1')$ and $(P2)$ read

$$
\begin{array}{llllll}
x_1 & + & x_2 & & + & x_4 & \geq & b_1 & , \\
x_1 & & & + & x_3 & + & x_4 & \geq & b_2 & , & (P1'') \\
& & x_2 & + & x_3 & + & x_4 & \geq & b_3 & , \\
x_1 & + & x_2 & + & x_3 & & & = & b_4 & , & (P2')
\end{array}
$$

and the constraints $(D1)$, $(D2)$ are given by

$$
\begin{array}{llllll}
y_1 & + & y_2 & & + & y_4 & \leq & 0 & , \\
y_1 & & & + & y_3 & + & y_4 & \leq & 0 & , & (D1') \\
& & y_2 & + & y_3 & + & y_4 & \leq & 0 & , \\
y_1 & + & y_2 & + & y_3 & + & y_4 & \leq & 1 & . & (D2')
\end{array}
$$

From $(D1')$ we infer that

$$2(y_1 + y_2 + y_3) + 3y_4 \leq 0 ,$$

hence

$$y_4 \leq -\frac{2}{3}(y_1 + y_2 + y_3) .$$

This implies that

$$b_1 y_1 + b_2 y_2 + b_3 y_3 + b_4 y_4 \leq (b_1 - \frac{2}{3}b_4)y_1 + (b_2 - \frac{2}{3}b_4)y_2 + (b_3 - \frac{2}{3}b_4)y_3 \leq 0$$

for all $y_1 \geq 0$, $y_2 \geq 0$, $y_3 \geq 0$, if $b_4 \geq \frac{3}{2}b_i$ for $i = 1, 2, 3$ which is then a sufficient condition for the core $c(v_t)$ of v_t to be non-empty. If we choose

$$x_1 = x_2 = x_3 = \frac{1}{3}b_4 ,$$

then

$$x_1 + x_2 = \frac{2}{3}b_4 \geq b_1 ,$$

$$x_1 + x_3 = \frac{2}{3}b_4 \geq b_2 ,$$

$$x_2 + x_3 = \frac{2}{3}b_4 \geq b_3 ,$$

$$x_1 + x_2 + x_3 = b_4 .$$

Hence $(\frac{1}{3}b_4, \frac{1}{3}b_4, \frac{1}{3}b_4)$ is in the core $c(v_t)$ of v_t. One can easily see that under the condition

$$b_4 \geq \frac{3}{2}b_i \qquad \text{for} \qquad i = 1, 2, 3$$

all the vectors $(x_1, x_2, x_3)^T \in \mathbb{R}^3$ with

$$x_i \geq 0 \,, \ i = 1, 2, 3 \,, \ \sum_{i=1}^{3} x_i = b_4 \qquad \text{and}$$

$$x_1 \geq \frac{1}{2}(b_1 + b_2 - b_3) \,,$$

$$x_2 \geq \frac{1}{2}(b_1 - b_2 + b_3) \,,$$

$$x_3 \geq \frac{1}{2}(-b_1 + b_2 + b_3) \,,$$

are in the core $c(v_t)$ of v_t.
In the case $n = 4$ the constraints $(D1)$ read

$$
\begin{array}{llll}
y_1 + y_2 + y_3 & & + y_7 + y_8 + y_9 & + y_{11} \leq 0 \,, \\
y_1 & + y_4 + y_5 & + y_7 + y_8 & + y_{10} + y_{11} \leq 0 \,, \\
y_2 & + y_4 & + y_6 + y_7 & + y_9 + y_{10} + y_{11} \leq 0 \,, \\
y_3 & + y_5 + y_6 & + y_8 + y_9 + y_{10} + y_{11} \leq 0 \,,
\end{array}
$$

which implies

$$2(y_1 + y_2 + y_3 + y_4 + y_5 + y_6) + 3(y_7 + y_8 + y_9 + y_{10}) + 4y_{11} \leq 0$$

and in turn

$$y_{11} \leq -\frac{1}{2}(y_1 + y_2 + y_3 + y_4 + y_5 + y_6) - \frac{3}{4}(y_7 + y_8 + y_9 + y_{10}) \,.$$

Hence

$$\sum_{i=1}^{11} b_i y_i \leq \sum_{i=1}^{6}(b_i - \frac{1}{2}b_{11})y_i + \sum_{i=7}^{10}(b_i - \frac{3}{4}b_{11})y_i \leq 0$$

for all $y_i \geq 0$, $i = 1, \ldots, 10$, if

$$b_{11} \geq \begin{cases} 2b_i & \text{for } i = 1, \ldots, 6 \,, \\ \frac{4}{3}b_i & \text{for } i = 7, \ldots, 10 \,. \end{cases}$$

These conditions are therefore sufficient for the core $c(v_t)$ of v_t to be non-empty.

If we choose $x_1 = x_2 = x_3 = x_4 = \frac{1}{4}b_{11}$, then it follows that

$$
\begin{aligned}
x_1 + x_2 &= \tfrac{1}{2}b_{11} \geq b_1 \ , \\
x_1 \phantom{{}+{}} + x_3 &= \tfrac{1}{2}b_{11} \geq b_2 \ , \\
x_1 \phantom{{}+ x_3{}} + x_4 &= \tfrac{1}{2}b_{11} \geq b_3 \ , \\
x_2 + x_3 &= \tfrac{1}{2}b_{11} \geq b_4 \ , \\
x_2 \phantom{{}+{}} + x_4 &= \tfrac{1}{2}b_{11} \geq b_5 \ , \\
x_3 + x_4 &= \tfrac{1}{2}b_{11} \geq b_6 \ , \\
x_1 + x_2 + x_3 &= \tfrac{3}{4}b_{11} \geq b_7 \ , \\
x_1 + x_2 \phantom{{}+ x_3{}} + x_4 &= \tfrac{3}{4}b_{11} \geq b_8 \ , \\
x_1 \phantom{{}+ x_2{}} + x_3 + x_4 &= \tfrac{3}{4}b_{11} \geq b_9 \ , \\
x_2 + x_3 + x_4 &= \tfrac{3}{4}b_{11} \geq b_{10} \ .
\end{aligned}
$$

Hence $(\frac{1}{4}b_{11}, \frac{1}{4}b_{11}, \frac{1}{4}b_{11}, \frac{1}{4}b_{11})$ is in the core $c(v_t)$ of v_t. For a general $n \geq 3$ one derives as sufficient conditions for the core $c(v_t)$ of v_t to be non-empty:

$$
b_i - \frac{2}{n}b_{2^n - n - 1} \leq 0 \text{ for } i = 1, ..., \binom{n}{2} \ ,
$$

$$
b_i - \frac{3}{n}b_{2^n - n - 1} \leq 0 \text{ for } i = \binom{n}{2} + 1, ..., \binom{n}{2} + \binom{n}{3} \ ,
$$

...

$$
b_i - \frac{n-1}{n}b_{2^n - n - 1} \leq 0
$$

$$
\text{for } i = \binom{n}{2} + \cdots + \binom{n}{n-2} + 1, ..., \binom{n}{2} + \cdots + \binom{n}{n-1}
$$

Under these conditions it then follows that

$$
\underbrace{(\frac{1}{n}b_{2^n - n - 1}, \dots, \frac{1}{n}b_{2^n - n - 1})}_{n - times} \text{ is in the core } c(v_t) \text{ of } v_t.
$$

Let $K_r^i \subseteq \mathcal{N}$ for $i = 1, \dots, \binom{n}{r}$ be the $\binom{n}{r}$ subsets of $\mathcal{N}$ with r elements. Then the constraints $(P1)$ can be written in the form

$$
\sum_{k \in K_r^i} x_k \geq v_t(K_r^i) \ , \ i = 1, \dots, \binom{n}{r} \ , \ r = 2, \dots, n - 1 \ .
$$

This implies for every $r = 2, \dots, n - 1$ $\frac{r}{n}\binom{n}{r}\sum_{k=1}^{n} x_k \geq \sum_{i=1}^{\binom{n}{r}} v_t(K_r^i)$ and further

$\frac{r}{n}v_t(\mathcal{N}) \geq \frac{1}{\binom{n}{r}}\sum_{i=1}^{\binom{n}{r}} v_t(K_r^i)$ for $r = 2, \dots, n - 1$ as necessary conditions for the core $c(v_t)$ of v_t to be non-empty. If for every $r = 2,\ldots,n-1$ all $v_t(K_r^i)$ for $i = 1,\ldots, \binom{n}{r}$ are equal, then the last conditions imply

$$\frac{r}{n}v_t(\mathcal{N}) \geq v_t(K_r^i) \quad \text{for all } i = 1,\ldots, \binom{n}{r} \text{ and } \quad r = 2,\ldots,(n-1)$$

which are sufficient conditions for the core $c(v_t)$ of v_t to be non-empty. Let us again consider the case $n = 3$. In this case the condition

$$v_t(\mathcal{N}) \geq \frac{1}{2}(v_t(K_2^1) + v_t(K_2^2) + v_t(K_2^3))$$

is necessary for the core $c(v_t)$ of v_t to be non-empty. Now let us assume in addition that

$$v_t(K_2^1) + v_t(K_2^2) - v_t(K_2^3) \geq 0 \ ,$$
$$v_t(K_2^1) - v_t(K_2^2) + v_t(K_2^3) \geq 0 \ ,$$
$$-v_t(K_2^1) + v_t(K_2^2) + v_t(K_2^3) \geq 0 \ .$$

Then we put

$$\varepsilon = v_t(\mathcal{N}) - \frac{1}{2}(v_t(K_2^1) + v_t(K_2^2) + v_t(K_2^3))$$

and define

$$x_1 = \frac{1}{2}(v_t(K_2^1) + v_t(K_2^2) - v_t(K_2^3)) + \frac{\varepsilon}{3} \ ,$$
$$x_2 = \frac{1}{2}(v_t(K_2^1) - v_t(K_2^2) + v_t(K_2^3)) + \frac{\varepsilon}{3} \ ,$$
$$x_3 = \frac{1}{2}(-v_t(K_2^1) + v_t(K_2^2) + v_t(K_2^3)) + \frac{\varepsilon}{3} \ .$$

From these definitions we infer

$$x_1 \geq 0 \ , \ x_2 \geq 0 \ , \ x_3 \geq 0 \ ,$$
$$x_1 + x_2 = v_t(K_2^1) + \frac{2}{3}\varepsilon \geq v_t(K_2^1) \ ,$$
$$x_1 + x_3 = v_t(K_2^2) + \frac{2}{3}\varepsilon \geq v_t(K_2^2) \ ,$$
$$x_2 + x_3 = v_t(K_2^3) + \frac{2}{3}\varepsilon \geq v_t(K_2^3) \ ,$$

and

$$x_1 + x_2 + x_3 = \frac{1}{2}(v_t(K_2^1) + v_t(K_2^2) + v_t(K_2^3) + \varepsilon) = v_t(\mathcal{N}) \ .$$

Hence (x_1, x_2, x_3) is in the core $c(v_t)$ of v_t.

3.4.5 A Second Cooperative Treatment

Now let us come back to the problem of minimizing (3.28) subject to (3.26), (3.27).

Let us define, for every non-empty subset S of $\mathcal{N} = \{1, \ldots, r\}$,

$$c_j(S) = \sum_{i \in S} c_{ij} \quad \text{for } j = 1, \ldots, r$$

and

$$b(S) = \sum_{i \in S} b_i .$$

We assume that, for all non-empty $S \subseteq \mathcal{N}$,

$$(A) \begin{cases} c_{ij} \geq 0 & \text{for all } i, j = 1, \ldots, r , \\ c_{ii} > 0 & \text{and} \\ b_i > 0 & \text{for } i = 1, \ldots, r . \end{cases}$$

For every non-empty $S \subseteq \mathcal{N}$ we now consider the problem of minimizing

$$\sum_{j \in \mathcal{N}} x_j$$

subject to

$$x_j \geq 0 \qquad \text{for } j = 1, \ldots, r \tag{3.39}$$

and

$$\sum_{j=1}^{r} c_j(S) x_j \geq b(S) . \tag{3.26$_S$}$$

According to the assumption (A) the set $V(S)$ of all vectors $x \in \mathbb{R}^r$ which satisfy (3.27) and (3.26$_S$) is non-empty and, since $\sum_{j \in S} x_j \geq 0$ for all $x \in V(S)$, there exists some $x(S) \in V(S)$ with

$$\sum_{j \in S} x_j(S) = \min\{\sum_{j \in S} x_j \mid x \in V(S)\} =: v(S) .$$

This function $v : 2^{\mathcal{N}} \to \mathbb{R}_+$ can be interpreted as the payoff function of a cooperative r-person game, if we define $v(\phi) = 0$. The subsets $S \subseteq \mathcal{N}$ can be interpreted as coalition which are built by the players by adding their constraints in (3.26) and minimizing $\sum_{j \in \mathcal{N}} x_j$.

The value $v(S)$, for every non-empty $S \subseteq \mathcal{N}$, can be determined explicitly. For that purpose we consider the dual problem which consists of maximizing $b(S)y$ subject to

$$c_j(S)y \leq 1 \qquad \text{for all } j \in \mathcal{N} \text{ and } y \geq 0 .$$

This problem has a solution, namely

$$y(S) = \max\{ \; \frac{1}{c_j(S)} \; | \; j \in \mathcal{N} \; , \; c_j(S) > 0 \; \}$$

and it is

$$v(S) = b(S) \cdot y(S) \; .$$

The question now arises what could be an incentive for the players to join in a grand coalition. For that purpose we divide the minimum cost $v(\mathcal{N})$ of the grand coalition $\mathcal{N}$ into r shares $x_i \geq 0$, $i = 1, \ldots, r$, i.e.,

$$\sum_{i=1}^{r} x_i = v(\mathcal{N}) \; . \tag{$C_{\mathcal{N}}$}$$

If this can be done in such a way that for every coalition $S \subseteq \mathcal{N}$ we have

$$\sum_{i \in S} x_i \leq v(S) \; , \tag{C_S}$$

then there is no incentive for them to form coalitions other than the grand one. In such a case we call the grand coalition stable.
Now we can prove the following

Theorem 3.5. *If the condition* (A) *is satisfied, then the grand coalition is stable.*

For the proof of this theorem we need the following lemma whose proof is modelled along the lines of the proof of *Theorem 1* in [24].

Lemma 3.1. *If the condition* (A) *is satisfied, then for every collection* $\mathcal{B} \subseteq 2^{\mathcal{N}}$ *such that there exist weights* $\gamma_S \geq 0$ *for* $S \in \mathcal{B}$ *with*

$$\sum_{\substack{S \in \mathcal{B} \\ i \in S}} \gamma_S = 1 \quad \text{for all} \quad i \in \mathcal{N} \tag{B_1}$$

it follows that

$$v(\mathcal{N}) \leq \sum_{S \in \mathcal{B}} \gamma_S v(S) \; . \tag{B_2}$$

Proof. Let $\mathcal{B} \subseteq 2^{\mathcal{N}}$ a collection as required. Then it follows that

$$\sum_{S \in \mathcal{B}} \gamma_S b(S) = \sum_{S \in \mathcal{B}} \sum_{i \in S} \gamma_S b_i = \sum_{i \in \mathcal{N}} \left(\sum_{\substack{S \in \mathcal{B} \\ i \in S}} \gamma_S \right) b_i = b(\mathcal{N}) .$$

Now let, for every $S \in \mathcal{B}$,

$$v(S) = x_1(S) + \ldots + x_r(S)$$

where $(x_1(S), \ldots, x_r(S))$ satisfies (3.27), (3.26)$_S$. Then

$$\sum_{S \in \mathcal{B}} \gamma_S v(S) = \sum_{S \in \mathcal{B}} \gamma_S \sum_{j=1}^{r} x_j(S)$$

$$= \sum_{j=1}^{r} \left(\sum_{S \in \mathcal{B}} \gamma_S x_j(S) \right) = \sum_{j=1}^{r} \hat{x}_j$$

where

$$\hat{x}_j = \sum_{S \in \mathcal{B}} \gamma_S x_j(S) \quad \text{for} \quad j = 1, \ldots, r .$$

Now we have

$$\sum_{j=1}^{r} c_j(S) \left(\sum_{S \in \mathcal{B}} \gamma_S x_j(S) \right) = \sum_{S \in \mathcal{B}} \gamma_S \left(\sum_{j=1}^{r} c_j(S) x_j(S) \right)$$

$$\geq \sum_{S \in \mathcal{B}} \gamma_S b(S)$$

which implies

$$\sum_{j=1}^{r} c_j(S) \hat{x}_j \geq b(N) , \quad \text{hence} \quad \sum_{j=1}^{r} c_j(\mathcal{N}) \hat{x}_j \geq b(\mathcal{N}) .$$

Since $\hat{x}_j \geq 0$ for $j = 1, \ldots, r$, it follows that

$$v(N) \leq \sum_{j=1}^{r} \hat{x}_j = \sum_{j=1}^{r} \sum_{S \in \mathcal{B}} \gamma_S x_j(S)$$

$$= \sum_{S \in \mathcal{B}} \gamma_S \sum_{j=1}^{r} x_j(S) = \sum_{S \in \mathcal{B}} \gamma_S v(S) .$$

This completes the proof.

$\square$

The proof of *Theorem 3.5* can be given in the same way as the proof of *Theorem 8.4* in Chapter II of [3] and goes as follows:

Proof. Let us consider the problem of maximizing $\sum_{j \in \mathcal{N}} x_j$ subject to (3.27), (C_S) for all $S \subseteq \mathcal{N}$. The problem which is dual to this consists of minimizing $\sum_{S \in \mathcal{B}} v(S)\gamma_S$, $\mathcal{B} = \{S \subseteq \mathcal{N} \mid S \neq \emptyset\}$. under the conditions

$$\gamma_S \geq 0 \quad \text{for all} \quad S \in \mathcal{B} \tag{D_1}$$

and

$$\sum_{\substack{S \in \mathcal{B} \\ i \in S}} \gamma_S \geq 1 \quad \text{for} \quad i = 1, \ldots, r . \tag{D_2}$$

The problem is solvable and

$$\max\{\sum_{i \in \mathcal{N}} x_i \mid x \in \mathbb{R}^n \text{ satisfies } (3.27), (C_S) \text{ for all } S \subseteq \mathcal{N}\} \leq v(\mathcal{N}) .$$

Therefore the dual problem is also solvable and

$$\min\{\sum_{S \in \mathcal{B}} v(S)\gamma_S \mid (\gamma_S)_{S \in \mathcal{B}} \text{ satisfies } (D_1), (D_2)\}$$
$$= \max\{\sum_{i \in \mathcal{N}} x_i \mid x \in \mathbb{R}^n \text{ satisfies } (3.27), (C_S) \text{ for all } S \subseteq \mathcal{N}\} \leq v(\mathcal{N}) .$$

For a solution $(\hat{\gamma}_S)_{S \in \mathcal{B}}$ of the dual problem we can assume that

$$\sum_{\substack{S \in \mathcal{B} \\ i \in S}} \hat{\gamma}_S = 1 \quad \text{for} \quad i = 1, \ldots, r .$$

Therefore *Lemma 3.1* implies

$$v(\mathcal{N}) \leq \min\{\sum_{S \in \mathcal{B}} v(S)\gamma_S \mid (\gamma_S)_{S \in \mathcal{B}} \text{ satisfies } (D_1) , (D_2)\}$$

hence

$$v(\mathcal{N}) = \max\{\sum_{i \in \mathcal{N}} x_i \mid x \in \mathbb{R}^n \text{ satisfies } (3.45) , (C_S) \text{ for all } S \subseteq \mathcal{N}\}$$

which shows that the grand coalition is stable and completes the proof of *Theorem 3.5*.

$\square$

In order to derive further sufficient and necessary conditions for the grand coalition to be stable we denote the $\binom{r}{k}$ subsets of $\mathcal{N}$ which have k elements by S_k^l, $l = 1, \ldots, \binom{r}{k}$.

Then we have the conditions

$$\sum_{i \in S_k^l} x_i \leq v(S_k^l) \qquad\qquad (C_r^k)$$

to be satisfied for $l = 1, \ldots, \binom{r}{k}$ and $k = 1, \ldots, r - 1$ together with

$$\sum_{i=1}^r x_i = v(\mathcal{N}) . \qquad\qquad (C_r^r)$$

Now let us assume that

$$\frac{k}{r} v(N) \leq \min_{l=1,\ldots,\binom{r}{k}} v(S_k^l) \qquad \text{for } k = 1, \ldots, r - 1 .$$

If we then define

$$x_i = \frac{1}{r} v(\mathcal{N}) \qquad \text{for } i = 1, \ldots, r$$

we obtain

$$\sum_{i \in S_k^l} x_i = \frac{k}{r} v(N) \leq v(S_k^l) \qquad \text{for all } l = 1, \ldots, \binom{r}{k} \text{ and } k = 1, \ldots, r - 1$$

and

$$\sum_{i=1}^r x_i = v(\mathcal{N}) ,$$

i.e. , $\{x_1, \ldots, x_r\}$ is a division of $v(\mathcal{N})$ of the kind we are looking for. On using the fact that

$$\sum_{l=1}^{\binom{r}{k}} \sum_{i \in S_k^l} x_i = \binom{r}{k} \cdot \frac{k}{r} \sum_{i=1}^r x_i \qquad \text{for all } k = 1, \ldots, r$$

one can see that

$$\frac{k}{r} v(\mathcal{N}) \leq \frac{1}{\binom{r}{k}} \sum_{l=1}^{\binom{r}{k}} v(S_k^l) \qquad \text{for all } k = 1, \ldots, r - 1$$

are necessary conditions for the existence of a division $\{x_1, \ldots, x_r\}$ of $v(\mathcal{N})$
with

$$\sum_{i \in S} x_i \leq v(S) \text{ for all } S \subseteq \mathcal{N} \ .$$

In order to obtain further necessary conditions we enumerate the sets S_k^l for
and $k = 1, \ldots, r$ such that

$$S_k^l \cap S_{r-k}^{\binom{r}{k}-l+1} = \emptyset \ .$$

Then it follows from (C_r^k) that

$$v(\mathcal{N}) \leq v(S_k^l) + v(S_{r-k}^{\binom{r}{k}-l+1}) \ , \qquad l = 1, \ldots, \binom{r}{k} \ .$$

For $r = 3$ we get as necessary conditions for the existence of a division
$\{x_1, x_2, x_3\}$ of $v(\mathcal{N})$ with

$$0 \leq x_i \leq v(S_1^i) \qquad \text{for } i = 1, 2, 3, \tag{C_3^1}$$

$$\begin{array}{rl} x_1 + x_2 & \leq v(S_2^1) \ , \\ x_1 \quad + x_3 & \leq v(S_2^2) \ , \\ x_2 + x_3 & \leq v(S_2^3) \ , \end{array} \tag{C_3^2}$$

the conditions

$$v(\mathcal{N}) \leq \sum_{l=1}^{3} v(S_1^l) \ ,$$

$$2v(\mathcal{N}) \leq \sum_{l=1}^{3} v(S_2^l) \ ,$$

$$v(\mathcal{N}) \leq v(S_1^l) + v(S_2^{4-l}) \ , \ l = 1, 2, 3,$$

$$(v(\mathcal{N}) \leq v(S_2^l) + v(S_1^{4-l}) \ , \ l = 1, 2, 3) \ .$$

Now let

$$(*) \qquad \begin{cases} 2v(\mathcal{N}) \leq \displaystyle\sum_{l=1}^{3} v(S_2^l) \\ \text{and} \\ v(\mathcal{N}) \geq v(S_2^l) \qquad \text{for } l = 1, 2, 3 \end{cases}$$

be satisfied.
Then we choose $b_1, b_2, b_3 \in \mathbb{R}$ such that

$$0 \leq b_i \leq v(S_2^i) \qquad \text{for } i = 1, 2, 3$$

and

$$b_1 + b_2 + b_3 = 2v(\mathcal{N}) .$$

If we then define

$$
\begin{array}{ll}
x_1 = \frac{1}{2}(b_1 + b_2) - \frac{1}{2}b_3 , & \\
x_2 = \frac{1}{2}(b_1 + b_3) - \frac{1}{2}b_2 , & (**) \\
x_3 = \frac{1}{2}(b_2 + b_3) - \frac{1}{2}b_1 , &
\end{array}
$$

we obtain

$$
\begin{array}{l}
x_1 = v(\mathcal{N}) - b_3 \geq v(\mathcal{N}) - v(S_2^3) \geq 0 , \\
x_2 = v(\mathcal{N}) - b_2 \geq v(\mathcal{N}) - v(S_2^2) \geq 0 , \\
x_3 = v(\mathcal{N}) - b_1 \geq v(\mathcal{N}) - v(S_2^1) \geq 0 ,
\end{array}
$$

$$
\begin{array}{lll}
x_1 + x_2 & = b_1 \leq v(S_2^1) , & \\
x_1 + x_3 & = b_2 \leq v(S_2^2) , & (C_3^2) \\
x_2 + x_3 & = b_3 \leq v(S_2^3) , &
\end{array}
$$

and

$$x_1 + x_2 + x_3 = \frac{1}{2}(b_1 + b_2 + b_3) = v(\mathcal{N}) . \qquad (C_3^3)$$

Therefore x_1, x_2, x_3 defined by $(**)$ satisfy (C_3^2) and (C_3^3).

Insertion of x_1, x_2, x_3 into (C_3^1) leads to the conditions

$$
\begin{array}{l}
\frac{1}{2}(b_1 + b_2) - \frac{1}{2}b_3 \leq v(S_1^1) , \\
\frac{1}{2}(b_1 + b_3) - \frac{1}{2}b_2 \leq v(S_1^2) , \\
\frac{1}{2}(b_2 + b_3) - \frac{1}{2}b_1 \leq v(S_1^3) ,
\end{array}
$$

which together with $(*)$ are sufficient for the existence of a division $\{x_1, x_2, x_3\}$ of $v(\mathcal{N})$ with (C_3^1), (C_3^2). The implication $(B_1) \Rightarrow (B_2)$ is also necessary for the existence of some $x \in \mathbb{R}^n$ with (3.27), $(C_\mathcal{N})$ and (C_S) for all $S \subseteq \mathcal{N}$. For, if such an $x \in \mathbb{R}^n$ is given, then, for every collection $\mathcal{B} \subseteq 2^\mathcal{N}$ such that there exist weights $\gamma_S \geq 0$ for $S \in \mathcal{B}$ with (B_1), it follows that

$$v(\mathcal{N}) = \sum_{j=1}^n x_j = \sum_{j=1}^n \Big(\sum_{\substack{S \in \mathcal{B} \\ j \in S}} \gamma_S \Big) x_j = \sum_{S \in \mathcal{B}} \gamma_S \sum_{j \in S} x_j \leq \sum_{S \in \mathcal{B}} \gamma_S v(S) .$$

In order to calculate an $x \in \mathbb{R}^n$ with (3.27), $(C_\mathcal{N})$, (C_S) for all $S \subseteq \mathcal{N}$ one can solve the problem of maximizing $\sum\limits_{i \in \mathcal{N}} x_i$ subject to (3.27), (C_S) for all $S \subseteq \mathcal{N}$. This can be done with the aid of the simplex method which can be performed without the initializing phase by introducing slack variables

$$z(S) \geq 0 \text{ for all } S \subseteq \mathcal{N} \quad S \neq \emptyset , \qquad (***)$$

and rewriting the inequalities (C_S) in the form

$$z(S) + \sum_{j \in S} x_j = v(S) \qquad (\tilde{C}_S)$$

for $S \subseteq \mathcal{N}$, $S \neq \emptyset$.

Then one has to maximize

$$\sum_{\substack{S \subseteq \mathcal{N} \\ S \neq \emptyset}} 0 \cdot z(S) + \sum_{j \in \mathcal{N}} x_j \qquad (**)$$

subject to $(***)$, (3.27) and $(\tilde{C}_S)$ for all $S \subseteq \mathcal{N}$, $S \neq \emptyset$. Let us demonstrate this by an example: As system (3.26) we consider

$$
\begin{aligned}
x_1 \phantom{{}+{}} + 0.8x_2 + 0.1x_3 &\geq 0.2 , \\
0.2x_1 + x_2 + 0.8x_3 &\geq 0.2 , \\
0.1x_1 + 0.5x_2 + x_3 &\geq 0.2 .
\end{aligned}
$$

In this case we get

$$v(S_1^1) = v(S_1^2) = v(S_1^3) = 0.2 ,$$

$$v(S_2^1) = v(S_2^3) = 0.\overline{2}\ldots , \quad v(S_2^2) = 0.3076923 , \quad v(\mathcal{N}) = 0.2608696$$

and the initial simplex tableau reads as follows:

	$v(S)$	$-x_1$	$-x_2$	$-x_3$
$z(S_1^1)$	0.2	1	0	0
$z(S_1^2)$	0.2	0	1	0
$z(S_1^3)$	0.2	0	0	1
$z(S_2^1)$	$0.\overline{2}\ldots$	1	1	0
$z(S_2^2)$	0.3076923	1	0	1
$z(S_2^3)$	$0.\overline{2}\ldots$	0	1	1
$z(N)$	0.2608696	1	1	1
$\sum_{i=1}^{3} x_i$	0	-1	-1	-1

After three simplex steps we arrive at the tableau.

	$v(S)$	$-z(S_1^1)$	$-z(S_2^1)$	$-z(\mathcal{N})$
x_1	0.2	1	0	0
$z(S_1^2)$	$0.\overline{7}\dots$	1	-1	0
$z(S_1^3)$	0.1613526	0	1	-1
x_2	$0.0\overline{2}\dots$	-1	1	0
$z(S_2^2)$	0.0690449	-1	0	-1
$z(S_2^3)$	0.1613526	1	0	-1
x_3	0.0386474	0	-1	1
$\sum_{i=1}^{3} x_i$	0.2608696	0	0	1

A solution of the problem of maximizing $(**)$ subject to $(***)$, (3.27) and $(\tilde{C}_S)$ for all $S \subseteq \mathcal{N}$, $S \neq \emptyset$ is given by the first column of the tableau together with $z(S_1^1) = z(S_2^1) = z(\mathcal{N}) = 0$. Further we obtain $\sum_{i=1}^{3} x_i = 0.2608696 = v(\mathcal{N})$.

3.5 A Dynamical Method for Finding a Nash Equilibrium

3.5.1 The Goal-Cost-Game

We consider n players $P_1, \dots, P_n$ that persue n goals $\hat{x}_i \in \mathbb{R}^{n_i}$ for $i = 1, \dots, n$. Every player P_i has m_i actions $A_{i1}, \dots, A_{im_i}$ available which he can take in order to achieve his goal $\hat{x}_i$. Let us assume that action A_{ik} requires costs in the amount of $u_{ik} \geq 0$ which can be chosen by P_i. Let us further assume that every player P_i can spend at most costs in the amount of $C_i^* > 0$ which leads to the constraints

$$\sum_{k=1}^{m_i} u_{ik} \leq C_i^* \quad \text{for} \quad i = 1, \dots, n . \tag{3.40}$$

Finally, we assume that the goal vector of player P_i, if every player P_j has chosen his cost vector $u_j = (u_{j_1}, \dots, u_{jm_j})$ for $j = 1, \dots, n$, is given by $x_i(u_1, \dots, u_n)$ where $x_i : \mathbb{R}^{m_1} \times \mathbb{R}^{m_n} \to \mathbb{R}^{n_i}$ is a given vector function. That means that the goal vector of each player depends on the cost vectors of all players.

Then the problem the players have to solve consists of finding n cost vectors $u_i \in \mathbb{R}_+^{m_i} = \{u \in \mathbb{R}^{m_i} \mid u \geq \Theta_{m_i}\}$ for $i = 1, \dots, n$ which satisfy (3.40) and solve the system of equations given by

$$x_i(u_1, \dots, u_n) = \hat{x}_i \text{ for } i = 1, \dots, n . \tag{3.41}$$

For every player P_i we define a cost function $\phi_i : \mathbb{R}^{\sum\limits_{i=1}^{n} m_i} \to \mathbb{R}$ by

$$\phi_i(u_1, \ldots, u_n) = \|x_i(u_1, \ldots, u_n) - \hat{x}_i\|_2^2$$

for $i = 1, \ldots, n$ where $\| \cdot \|$ denotes the Euclidean norm in $\mathbb{R}^{n_i}$.

Instead of finding a solution $u \in \mathbb{R}_+^{\sum\limits_{i=1}^{n} m_i}$ of (3.41) which satisfies (3.40) the players now try to find a Nash equilibrium, i.e., to find vectors $u_i^* \in \mathbb{R}^{m_i}$ with

$$u_i^* \geq \Theta_{m_i} \text{ and } \sum_{k=1}^{m_i} u_{ik}^* \leq C_i^* \tag{3.42}$$

$$\text{for all } i = 1, \ldots, n$$

such that

$$\begin{cases} \varphi_i(u_1^*, \ldots, u_n^*) \leq \varphi_i(u_i^*, \ldots, u_{i-1}^*, u_i, u_{i+1}^*, \ldots, u_n^*) \\ \text{for all } u_i \in \mathbb{R}^{m_i} \text{ with} \\ u_i \geq \Theta_{m_i} \text{ and } \sum_{k=1}^{m_i} u_{ik} \leq C_i^* \\ \text{for all } i = 1, \ldots, n \ . \end{cases} \tag{3.43}$$

In order to solve this problem we shall make use of

3.5.2 Necessary Conditions for a Nash Equilibrium

We assume that $x_i \in C^1(\mathbb{R}^m, \mathbb{R}^{n_i})$ for every $i = 1, \ldots, n$ where $m = \sum\limits_{i=1}^{n} m_j$. Then it follows that $\varphi_i \in C^1(\mathbb{R}^m, \mathbb{R})$ for every $i = 1, \ldots, n$. A necessary condition for a Nash equilibrium is then given by the well known multiplier rule:

For every $i \in \{1, \ldots, n\}$ there exist numbers $\lambda_i \geq 0$ and $\lambda_{ik} \geq 0$ for $k = 1, \ldots, m_i$ such that

$$\varphi_{iu_{ik}}(u_1^*, \ldots, u_n^*) = -\lambda_i + \lambda_{ik} \quad \text{for } k = 1, \ldots, m_i \ ,$$

$$\lambda_i \left(\sum_{k=1}^{m_i} u_{ik}^* - C_i^* \right) = 0$$

and

$$\lambda_{ik} u_{ik}^* = 0 \quad \text{for } k = 1, \ldots, m_i \ .$$

If every $\varphi_i = \varphi_i(u_1, \ldots, u_n)$, $i = 1, \ldots, n$ is convex, then the multiplier rule is also sufficient for a Nash equilibrium.

For, if $(u_1, \ldots, u_n) \in \mathbb{R}^m$ is given with

$$u_i \geq \Theta_{m_i} \quad \text{and} \quad \sum_{k=1}^{m_i} u_{ik} \leq C_i^* \quad \text{for all } i = 1, \ldots, n \,,$$

then it follows that, for every $i \in \{1, \ldots, n\}$,

$$\varphi_i(u_1^*, \ldots, u_{i-1}^*, u_i, u_{i+1}^*, \ldots, u_n^*) - \varphi_i(u_1^*, \ldots, u_n^*)$$

$$\geq \sum_{k=1}^{m_i} \varphi_{iu_{ik}}(u_1^*, \ldots, u_n^*)(u_{ik} - u_{ik}^*)$$

$$= \sum_{k=1}^{m_i} (-\lambda_i + \lambda_{ik})(u_{ik} - u_{ik}^*)$$

$$= -\lambda_i \sum_{k=1}^{m_i} u_{ik} + \lambda_i \sum_{k=1}^{m_i} u_{ik}^* + \sum_{k=1}^{m_i} \lambda_{ik} u_{ik}$$

$$= \lambda_i (C_i^* - \sum_{k=1}^{m_i} u_{ik}) + \sum_{k=1}^{m_i} \lambda_{ik} u_{ik}$$

$$\geq 0 \,.$$

This implies that $(u_1^*, \ldots, u_n^*)$ with (3.42) for $i = 1, \ldots, n$ is a Nash equilibrium, if it satisfies the multiplier rule. Now let $(u_1^*, \ldots, u_n^*)$ satisfy (3.42) for all $i = 1, \ldots, n$ and be a Nash equilibrium, hence satisfy the multiplier rule. Let, for some $i \in \{1, \ldots, n\}$,

$$\sum_{k=1}^{m_i} u_{ik}^* < C_i^* \,. \tag{3.44}$$

Then it follows that $\lambda_i = 0$ which implies

$$\varphi_{iu_{ik}}(u_i^*, \ldots, u_n^*) \geq 0 \quad \text{and} \quad u_{ik}^* \varphi_{iu_{ik}}(u_1^*, \ldots, u_n^*) = 0$$

for all $k = 1, \ldots, m_i$.

$$\tag{3.45}$$

Let, for some $i \in \{1, \ldots, n\}$,

$$\sum_{k=1}^{m_i} u_{ik}^* = C_i^* \,. \tag{3.46}$$

Then it follows that

$$\varphi_{iu_{ik}}(u_1^*, \ldots, u_n^*) \leq 0 \quad \text{for all } u_{ik}^* > 0 \,. \tag{3.47}$$

3.5.3 The Method

The two necessary conditions (3.45) and (3.47) for a Nash equilibrium give rise to an iterative procedure for finding it. By this procedure a sequence $(u_1(t), \ldots, u_n(t))_{t \in \mathbb{N}_0}$ with

$$u_i(t) \geq \Theta_{m_i} \quad \text{and} \quad \sum_{k=1}^{m_i} u_{ik}(t) \leq C_i^* \quad \text{for all} \quad i = 1, \ldots, n \qquad (3.48)$$

is constructed as follows:
Let $u(t) = (u_1(t), \ldots, u_n(t)) \in \mathbb{R}^m$ with (3.48) be given for some $t \in \mathbb{N}_0$ and let $i \in \{1, \ldots, n\}$ be such that

$$\sum_{k=1}^{m_i} u_{ik}(t) < C_i^* \qquad (3.49)$$

Then we define the set

$$K_i = \{ k \in \{1, \ldots, m_i\} \mid \varphi_{iu_{ik}}(u_1(t), \ldots, u_n(t)) < 0 \quad \text{or}$$
$$u_{ik}(t)\varphi_{iu_{ik}}(u_1(t), \ldots, u_n(t)) \neq 0 \} . \qquad (3.50)$$

If K_i is empty, then (3.45) is satisfied for $(u_1(t), \ldots, u_n(t))$ instead of $(u_1^*, \ldots, u_n^*)$. In this case we put $u_i(t + 1) = u_i(t)$. Otherwise we put

$$u_{ik}(t + 1) = u_{ik}(t) + \lambda_i h_{ik}(t)$$

with $\lambda_i > 0$ and

$$h_{ik}(t) = \begin{cases} -\varphi_{iu_{ik}}(u_1(t), \ldots, u_n(t)) & \text{, if } k \in K_i , \\ 0 & \text{, if } k \notin K_i . \end{cases} \qquad (3.51)$$

If $u_{ik}(t) = 0$ for some $k \in \{1, \ldots, m_i\}$, then it follows that $h_{ik}(t) \geq 0$ and hence $u_{ik}(t + 1) \geq 0$ for all $\lambda_i > 0$.
Therefore we define the set

$$K_{i1} = \{ k \in \{1, \ldots, m_i\} \mid u_{ik}(t) > 0 \quad \text{and} \quad h_{ik}(t) < 0 \} \qquad (3.52)$$

and put

$$\lambda_{i1} = \begin{cases} \min\{ \frac{u_{ik}(t)}{-h_{ik}(t)} \mid k \in K_{i1} \} & \text{, if } K_{i1} \neq \emptyset \\ +\infty & \text{, else .} \end{cases} \qquad (3.53)$$

Then it follows that $\lambda_{i1} > 0$ and

$$u_{ik}(t + 1) = u_{ik}(t) + \lambda_i h_{ik}(t) \geq 0$$
$$\text{for all} \quad k = 1, \ldots, m_i \quad \text{and all} \quad \lambda_i \in (0, \lambda_{i1}] . \qquad (3.54)$$

If we put

$$
\lambda_{i2} =
\begin{cases}
\dfrac{C_i^* - \sum\limits_{k=1}^{m_i} u_{ik}(t)}{\sum\limits_{k=1}^{m_i} h_{ik}(t)} & , \text{ if } \sum\limits_{k=1}^{m_i} h_{ik}(t) > 0 \\[4ex]
+\infty & , \text{ else },
\end{cases}
$$

then it follows that $\lambda_{i2} > 0$ and

$$
\sum_{k=1}^{m_i} u_{ik}(t+1) = \sum_{k=1}^{m_i} (u_{ik}(t) + \lambda_i h_{ik}(t)) \le C_i^* \text{ for all } \lambda_i \in (0, \lambda_{i2}] . \qquad (3.55)
$$

Now let $i \in \{1, \ldots, n\}$ be such that

$$
\sum_{k=1}^{m_i} u_{ik}(t) = C_i^* . \qquad (3.56)
$$

Then we define the set

$$
K_i = \{k \in \{1, \ldots, m_i\} \mid u_{ik}(t) > 0\} \text{ and } \varphi_{iu_{ik}}(u_1(t), \ldots, u_n(t)) > 0\} . \qquad (3.57)
$$

If K_i is empty, we put $u_i(t+1) = u_i(t)$, since (3.47) is satisfied for $(u_1(t), \ldots, u_n(t))$ instead of $(u_1^*, \ldots, u_n^*)$.
Otherwise we define

$$
u_{ik}(t+1) = u_{ik}(t) + \lambda_i h_{ik}(t)
$$

with $\lambda_i > 0$ and

$$
h_{ik}(t) =
\begin{cases}
-\varphi_{iu_{ik}}(u_1(t), \ldots, u_n(t)) & , \text{ if } k \in K_i , \\
0 & , \text{ if } k \notin K_i .
\end{cases}
\qquad (3.58)
$$

Then it follows that

$$
\sum_{k=1}^{m_i} u_{ik} = \sum_{k=1}^{m_i} (u_{ik}(t) + \lambda_i h_{ik}(t))
$$
$$
< \sum_{k=1}^{m_i} u_{ik}(t) = C_i^*
$$
$$
\text{for all } \lambda > 0 .
$$

If we define λ_{i1} by (3.53) with K_{i1} given by (3.52), then it follows that $\lambda_{i1} > 0$ and (3.54) is satisfied.
In both cases (3.49) and (3.56) it is true that

$$
\sum_{k=1}^{m_i} h_{ik}(t) \, \varphi_{iu_{ik}}(u_1(t), \ldots, u_n(t)) < 0 ,
$$

if the set K_i given by (3.50) and (3.57), respectively, are non-empty.
This implies that the vector $h_i(t)$ defined by (3.51) and (3.58), respectively,
determines a feasible descent direction. Therefore we put $\lambda_i^* = \min(\lambda_{i1}, \lambda_{i2})$
in the case (3.49) and $\lambda_i^* = \lambda_{i1}$ in the case (3.56), determine $\hat{\lambda}_i \in [0, \lambda_i^*]$ such
that

$$\varphi_i(u_1(t), \ldots, u_i(t) + \hat{\lambda}_i h_i(t), \ldots, u_n(t)) \leq$$
$$\varphi_i(u_1(t), \ldots, u_i(t) + \lambda_i h_i(t), \ldots, u_n(t))$$
$$\text{for all } \lambda_i \in [0, \lambda_i^*] ,$$

and put

$$u_i(t+1) = u_i(t) + \hat{\lambda}_i h_i(t) .$$

3.6 Evolution Matrix Games

3.6.1 Definition of the Game and Evolutionary Stability

We consider a population whose individuals have a finite number of strategies
$I_1, I_2, \ldots, I_n$ in order to survive in the struggle of life. Let $u_i \in [0,1]$, for
every $i = 1, \ldots, n$, be the probability for the strategy I_i to be chosen in the
population. Then the corresponding state of the population is defined by the
vector $u = (u_1, \ldots, u_n)$ where $\sum_{i=1}^{n} u_i = 1$. The space of all population states is
given by the simplex

$$\Delta = \{u = (u_1, \ldots, u_n) \mid 0 \leq u_1 \leq 1 , \sum_{i=1}^{n} u_i = 1\} .$$

Every vector $e_i = (0, \ldots, 0, 1_i, 0, \ldots, 0)$, $i = 1, \ldots, n$, denotes a so called pure
population state where all individuals choose the strategy I_i. All the other
states are called mixed states. If an individual that chooses strategy I_i meets
an individual that chooses strategy I_j, we assume that the I_i-individual is
given a payoff $a_{ij} \in \mathbb{R}$ by the I_j-individual. All the payoffs then form a matrix

$$A = (a_{ij})_{i,j=1,\ldots,n}$$

the so called payoff matrix which defines a matrix game.
The expected payoff of an I_i-individual in the population state $u \in \Delta$ is
defined by

$$\sum_{j=1}^{n} a_{ij} u_j = e_i A u^T .$$

If two population states $u, v \in \Delta$ are given, then the average payoff of v to u is defined by

$$\sum_{i,j=1}^{n} a_{ij} v_i u_j = v A u^T .$$

Definition 3.1. *A population state $u^* \in \Delta$ is called a Nash equilibrium, if*

$$u A u^{*^T} \leq u^* A u^{*^T} \quad \text{for all } u \in \Delta .$$

In words this means that a declination from u^* does not lead to a higher payoff.

Definition 3.2. *A Nash equilibrium $u^* \in \Delta$ is called evolutionary stable, if $u A u^{*^T} = u^* A u^{*^T}$ for some $u \in \Delta$ with $u \neq u^*$ implies that $u A u^T < u^* A u^T$.*

In words this means that, if a change from u^* to u leads to the same payoff, u cannot be a Nash equilibrium.

Let us demonstrate these definitions by an example. We consider a population with two strategies I_1 and I_2. Individuals that choose I_1 are called pigeons and those who choose I_2 are called hawks. If a pigeon meets a pigeon they menace each other without seriously fighting until one of them gives in. If a pigeon meets a hawk, it runs away and is not hurt. If two hawks meet each other, they fight until one of them is seriously hurt and has to give up or is dead. Let us assume that in each case the winner is given $V > 0$ points and the loser in a fight of hawks is given $-D$ points where $D > 0$. This leads to the payoff matrix

$$A = \begin{pmatrix} \frac{V}{2} & 0 \\ V & \frac{V-D}{2} \end{pmatrix} .$$

One can show that the pure population state $e_2 = (0, 1)$ where all individuals behave like hawks is evolutionary stable, if $V \geq D$ (Exercise). If $V < D$, then the pure population state $e_2 = (0, 1)$ is not even a Nash equilibrium. On the contrary we have

$$e_2 A e_2^T - v A e_2^T = \frac{V - D}{2}(1 - v_2) < 0 \text{ for all } v = (v_1, v_2) \in \Delta \text{ with } v_2 < 1 .$$

But also the pure population state $e_1 = (1, 0)$ is not a Nash equilibrium. In this case we have

$$e_1 A e_1^T - v A e_1^T = -\frac{V}{2}(1 - v_1) < 0 \text{ for all } v = (v_1, v_2) \in \Delta \text{ with } v_1 < 1 .$$

The case $V \geq D$ is a special case of the following situation:
Let, for some $k \in \{1, \dots, n\}$,

$$a_{kk} \geq a_{jk} \quad \text{for all } j = 1, \dots, n$$

and

$$a_{kk} = a_{jk} \quad \Rightarrow a_{ki} > a_{ji} \quad \text{for all } i \neq k .$$

(3.59)

Then, for every $u \in \Delta$, it follows that

$$u A e_k^T = \sum_{j=1}^{n} u_j a_{jk} \leq \left(\sum_{j=1}^{n} u_j \right) a_{kk} = a_{kk} = e_k A e_k^T .$$

Let $u A e_k^T = e_k A e_k^T$. Then

$$
\begin{aligned}
a_{jk} &= a_{kk} \quad \text{for all } j \text{ with } u_j > 0 \text{ , hence} \\
a_{ki} &> a_{ji} \quad \text{for all } j \text{ with } u_j > 0 \text{ and } i \neq k .
\end{aligned}
$$

(3.60)

This implies

$$
\begin{aligned}
e_k A u^T - u A u^T &= \sum_{i=1}^{n} a_{ki} u_i - \sum_{j=1}^{n} \sum_{i=1}^{n} a_{ji} u_j u_i \\
&= \sum_{j=1}^{n} \sum_{i=1}^{n} (a_{ki} - a_{ji}) u_j u_i = \sum_{u_j > 0} \sum_{i \neq k} (a_{ki} - a_{ji}) u_j u_i > 0
\end{aligned}
$$

(3.61)

and shows that $u^* = e_k$ is evolutionary stable. Evolutionary stability can be characterized by a condition which is useful for theoretical purposes. To derive this condition we start as follows:
Let $u, u^* \in \Delta$ be given with $u \neq u^*$ and let $\varepsilon \in (0, 1]$. Then we define $w_\varepsilon = (1 - \varepsilon)u^* + \varepsilon u$ and conclude that

$$E(w_\varepsilon, w_\varepsilon) = (1 - \varepsilon)E(u^*, w_\varepsilon) + \varepsilon E(u, w_\varepsilon)$$

where

$$E(u, v) = u A v^T \text{ for any } u, v \in \Delta .$$

From this we obtain the equivalence

$$E(w_\varepsilon, w_\varepsilon) < E(u^*, w_\varepsilon) \Leftrightarrow E(u, w_\varepsilon) < E(u^*, w_\varepsilon) .$$

(3.62)

Now let $u^* \in \Delta$ be evolutionary stable and let $u \in \Delta$ be chosen arbitrarily. Then we have

$$E(u, u^*) \leq E(u^*, u^*) .$$

1) Assume that $E(u, u^*) < E(u^*, u^*)$, $u \neq u^*$. Then there is a relatively open set $V_u \subseteq \Delta$ with $u^* \in V_u$ such that

$$E(u, v) < E(u^*, v) \text{ for all } v \in V_u \text{ with } v \neq u^* \ .$$

Now there exists some $\varepsilon_u > 0$, $\varepsilon_u \leq 1$ such that

$$w_\varepsilon = (1 - \varepsilon)u^* + \varepsilon u \in V_u \text{ for all } \varepsilon \in [0, \varepsilon_u] \ .$$

This implies

$$E(u, w_\varepsilon) < E(u^*, w_\varepsilon) \text{ for all } \varepsilon \in (0, \varepsilon_u] \ .$$

Using the above equivalence we obtain

$$E(w_\varepsilon, w_\varepsilon) < E(u^*, w_\varepsilon) \text{ for all } \varepsilon \in (0, \varepsilon_u] \ .$$

2) Assume that $E(u, u^*) = E(u^*, u^*)$, $u \neq u^*$. Then it follows that $E(u, u^*) > E(u, u)$ which implies

$$E(u, w_\varepsilon) < E(u^*, w_\varepsilon) \Leftrightarrow E(w_\varepsilon, w_\varepsilon) < E(u^*, w_\varepsilon)$$
$$\text{for all } \varepsilon \in (0, 1]. \tag{3.63}$$

Result: If $u^* \in \Delta$ is evolutionary stable, then, for every $u \in \Delta$ with $u \neq u^*$, there exists some $\varepsilon_u \in (0, 1]$ such that

$$E(w_\varepsilon, w_\varepsilon) < E(u^*, w_\varepsilon) \text{ for all } \varepsilon \in (0, \varepsilon_u] \tag{3.64}$$

where

$$w_\varepsilon = (1 - \varepsilon)u^* + \varepsilon u \ .$$

Conversely let $u^* \in \Delta$ be such that for every $u \in \Delta$ with $u \neq u^*$ there exists some $\varepsilon_u \in (0, 1]$ such that (3.64) is satisfied. Then it follows from the equivalence (3.62) that

$$E(u^*, w_\varepsilon) > E(u, w_\varepsilon) \tag{3.65}$$

and in turn for $\varepsilon \to 0$ that

$$E(u^*, u^*) \geq E(u, u^*) \ .$$

Let $E(u^*, u^*) = E(u, u^*)$. Then it follows from (3.64) that

$$(1 - \varepsilon)E(u^*, u^*) + \varepsilon E(u^*, u) > (1 - \varepsilon)E(u^*, u^*) + \varepsilon E(u, u)$$

which implies $E(u, u) < E(u^*, u)$.

Result: A population state $u^* \in \Delta$ is evolutionary stable, if and only if for every $u \in \Delta$ with $u \neq u^*$ there exists some $\varepsilon_u \in (0,1]$ such that the condition (3.64) is satisfied.

Now let $u^* \in \Delta$ be evolutionary stable and let

$$u_i^* > 0 \text{ for all } i = 1, \ldots, n \; . \tag{3.66}$$

Then it follows from

$$u^* A u^{*^T} = \sum_{i=1}^{n} u_i^* e_i A u^{*^T}$$

and

$$e_i A u^{*^T} \leq u^* A u^{*^T} \quad \text{for } i = 1, \ldots, n$$

that

$$e_i A u^{*^T} = u^* A u^{*^T} \quad \text{for } i = 1, \ldots, n$$

which implies

$$u A u^{*^T} = u^* A u^{*^T} \quad \text{for all } u \in \Delta$$

and in turn that

$$u A u^T < u^* A u^T \quad \text{for all } u \in \Delta \text{ with } u \neq u^* \; .$$

This shows that $u^* \in \Delta$ is the only evolutionary stable state. Let us define, for every $u \in \Delta$, a support set by

$$S(u) = \{i \mid u_i > 0\} \; .$$

Then it follows by the arguments given above that

$$e_i A u^{*^T} = u^* A u^{*^T} \quad \text{for all } i \in S(u^*) \tag{3.67}$$

which implies

$$u A u^{*^T} = u^* A u^{*^T} \quad \text{for all } u \in \Delta \text{ with } S(u) \subseteq S(u^*)$$

and in turn

$$u A u^T < u^* A u^T \quad \text{for all } u \in \Delta \text{ with } u \neq u^* \text{ and } S(u) \subseteq S(u^*) \; ,$$

if $u^* \in \Delta$ is an evolutionary stable state.

Now let $u \in \Delta$ be such that $S(u) \not\subseteq S(u^*)$. Then there exists some $i \in \{1, \ldots, n\}$ such that $u_i > 0$ and $u_i^* = 0$.

If
$$u_i \geq u_i^* \quad \text{for all } i = 1, \ldots, n \ ,$$

then it follows from
$$\sum_{i=1}^{n} u_i = \sum_{i=1}^{n} u_i^* = 1$$

that $u = u^*$ which is impossible.

Hence there exists some $i \in \{1, \ldots, n\}$ with $u_i < u_i^*$. If we define
$$\lambda = \min\{\frac{u_i^*}{u_i^* - u_i} \mid u_i < u_i^*\}$$

and put
$$v = u^* + \lambda(u - u^*) \ ,$$

then it follows that
$$v \in C = \{u \in \Delta \mid \exists i_1 \text{ with } u_{i_1} > 0 \text{ and } u_{i_1}^* = 0$$
$$\text{and } \exists i_2 \text{ with } u_{i_2} = 0\} \ .$$

Conversely, if $v \in C$ is given and we define, for any $\lambda \in (0, 1]$, $u = u^* + \lambda(v - u^*)$, then $u \in \Delta$ and $S(u) \not\subseteq S(u^*)$.

Now for every $v \in C$ there is some $\varepsilon_v \in (0, 1]$ such that
$$w_\varepsilon A w_\varepsilon^T < u^* A w_\varepsilon^T \text{ for all } \varepsilon \in (0, \varepsilon_v.]$$

where
$$w_\varepsilon = (1 - \varepsilon)u^* + \varepsilon v = u^* + \varepsilon(v - u^*) \ .$$

Since C is compact and ε_v, $v \in C$, can be chosen continuously, there exists some $\hat{\varepsilon} > 0$ with $\hat{\varepsilon} = \min_{v \in C} \varepsilon_v$ and therefore
$$w_\varepsilon A w_\varepsilon^T < u^* A w_\varepsilon^T \text{ for all } \varepsilon \in (0, \hat{\varepsilon}] \ .$$

If we define
$$\varepsilon^* = \frac{\hat{\varepsilon}}{\min_{v \in C} \|v - u^*\|_2} \ ,$$

then it follows that
$$u A u^T < u^* A u_\varepsilon^T \text{ for all } u \in \Delta$$
$$\text{with } S(u) \not\subseteq S(u^*) \text{ and } \|u - u^*\|_2 < \varepsilon^* \ .$$

Summarizing we obtain the

Result: If $u^* \in \Delta$ is evolutionary stable, then there exists some $\varepsilon^* > 0$ such that

$$uAu^T < u^*Au^T \text{ for all } u \in \Delta \text{ with } u \neq u^*$$
$$\text{and } \|u - u^*\|_2 < \varepsilon^* .$$

(3.68)

Conversely let $u^* \in \Delta$ and $\varepsilon^* > 0$ be given such that (3.68) is satisfied. If we then take any $u \in \Delta$ with $u \neq u^*$ and define, for $\varepsilon \in (0,1]$,

$$w_\varepsilon = (1 - \varepsilon)u^* + \varepsilon u ,$$

then $w_\varepsilon \in \Delta$, $w_\varepsilon \neq u^*$ and

$$\|w_\varepsilon - u^*\|_2 = \varepsilon\|u - u^*\|_2 < \varepsilon^*$$

for $\varepsilon < \min\left(1, \frac{\varepsilon^*}{\|u-u^*\|}\right) = \varepsilon_u \in (0,1)$ which implies

$$w_\varepsilon A w_\varepsilon^T < u^*Aw^T$$

and shows that (3.64) is satisfied and in turn that u^* is evolutionary stable.

Result: $u^* \in \Delta$ is evolutionary stable, if and only if there exists some $\varepsilon^* > 0$ such that (3.68) is satisfied.
The condition (3.68) says that an evolutionary stable state is locally the only evolutionary stable state.

3.6.2 A Dynamical Method for Finding an Evolutionary Stable State

Let us assume that

$$a_{ij} \geq 0 \text{ for all } i, j = 1, \ldots, n$$

and

$$uAu^T > 0 \text{ for all } u \in \Delta .$$

(3.69)

Then, starting with some $u^0 \in \Delta$, we define a sequence $(u^k)_{k \in \mathbb{N}_0}$ of population states by

$$u_i^{k+1} = \frac{e_i A(u^k)^T}{u^k A(u^k)^T} u_i^k \text{ for } i = 1, \ldots, n .$$

Obviously $u^k \in \Delta$ implies that $u^{k+1} \in \Delta$. If we define a map $f_A : \Delta \to \Delta$ by

$$f_A(u)_i = \frac{e_i A u^T}{u A u^T} u_i \text{ for } i = 1, \ldots, n \text{ and } u \in \Delta , \tag{3.70}$$

then

$$f_A(u^*) = u^* ,$$

if and only if

$$e_i A u^{*^T} = u^* A u^{*^T} \text{ for all } i \in S(u^*) . \tag{3.71}$$

Since in *Section 3.6.1* we have shown that this condition is necessary for $u^* \in \Delta$ to be evolutionary stable, it follows that $u^* \in \Delta$ is a fixed point of f_A, if u^* is evolutionary stable. This even holds true, if u^* is a Nash equilibrium, since the condition (3.71) is also necessary for u^* being a Nash equilibrium. This gives rise to the question under which condition a fixed point of f_A is a Nash equilibrium. A first answer to this question is

Lemma 3.2. *If $u^* \in \Delta$ is a fixed point of f_A and*

$$u_i^* > 0 \text{ for all } i = 1, \ldots, n , \tag{3.72}$$

then u^* is a Nash equilibrium.

Proof. $u^* \in \Delta$ is a fixed point of f_A, if and only if (3.71) holds true. Since $S(u^*) = \{1, \ldots, n\}$ this implies

$$u A u^{*^T} = u^* A u^{*^T} \quad \text{for all } u \in \Delta$$

which shows that u^* is a Nash equilibrium.

$\square$

A second answer to the above question is

Lemma 3.3. *If $u^* \in \Delta$ is an attractive fixed point (i.e. a fixed point which is an attractor), then $u^* \in \Delta$ is a Nash equilibrium.*

Proof. If u^* satisfies (3.72), the assertion follows from *Lemma 3.1*. If $S(u^*) \neq \{1, \ldots, n\}$, then it follows that (3.71) is satisfied. If we show that

$$e_i A u^{*^T} \leq u^* A u^{*^T} \quad \text{for all } i \in \{1, \ldots, n\} \setminus S(u^*) ,$$

then it follows that u^* is a Nash equilibrium.

Let us assume that, for some $k \in \{1, \ldots, n\} \setminus S(u^*)$, it is true that

$$e_k A u^{*^T} > u^* A u^{*^T} . \tag{3.73}$$

Since $g(u) = e_k A u^T - u A u^T$ is continuous, there is some $\varepsilon_1 > 0$ such that

$$e_k A u^T > u A u^T \text{ for all } u \in \Delta \text{ with } \|u - u^*\|_2 < \varepsilon_1 . \tag{3.74}$$

Since u^* is an attractor, there is some $\varepsilon_2 > 0$ such that

$$\lim_{t \to \infty} f_A^t(u) = u^* \text{ for all } u \in \Delta \text{ with } \|u - u^*\| < \varepsilon_2 . \tag{3.75}$$

This implies for every $v \in \Delta$ with $\|v - u^*\|_2 < \varepsilon$ the existence of some $T_\varepsilon \in \mathbb{N}$ such that

$$\|v(t) - u^*\| < \varepsilon \text{ for all } t \geq T_\varepsilon \text{ where } \varepsilon = \min(\varepsilon_1, \varepsilon_2)$$

and

$$v(t) = f_A^t(v) .$$

From (3.73) it follows that

$$v_k(t+1) > v_k(t) > 0 \text{ for all } t \in \mathbb{N} . \tag{3.76}$$

On the other hand (3.75) implies that

$$\lim_{t \to \infty} v_k(t) = u_k^* = 0 , \text{ since } k \in S(u^*) ,$$

which contradicts (3.76).
Hence the assumption (3.73) is false which completes the proof.

$\square$

The inversion of *Lemma 3.2* is in general false which can be shown by a counterexample (see [21]).
We can, however, prove

Theorem 3.6. *If a pure population state is evolutionary stable, then it is an asymptotically stable fixed point of f_A (3.70).*

Proof. Let e_k , for some $k \in \{1, \ldots, n\}$, be evolutionary stable.
Then by the second last result of *Section 3.6.2* there exists some $\varepsilon^* > 0$ such that

$$u A u^T < e_k A u^T \text{ for all } u \in \Delta \text{ with } u \neq e_k$$
$$\text{and } \|u - e_k\|_2 < \varepsilon^* .$$

Further e_k is a fixed point of f_A as shown above. In order to show that $\{e_k\}$ is asymptotically stable we verify the assumptions of *Theorem 1.3.* Let

$$U = \{u \in \Delta \mid \|u - e_k\|_2 < \varepsilon^*\} \, .$$

Further we define

$$G = \{u \in \Delta \mid |u_k - 1| < \frac{\varepsilon^*}{n}\} \, .$$

Then $G \subseteq \Delta$ is open in Δ, $e_k \in \Delta$ and for every $u = (u_1, \ldots, u_n) \in G$ it follows that

$$\sum_{\substack{i=1 \\ i \neq k}}^{n} u_i = 1 - u_k < \frac{\varepsilon^*}{n}$$

which implies

$$0 \leq u_i < \frac{\varepsilon^*}{n} \quad \text{for all } i \in \{1, \ldots, n\} \, , \; i \neq k \, ,$$

hence

$$\|u - e_k\|_2 < \frac{\varepsilon^*}{\sqrt{n}} < \varepsilon^* \, .$$

Therefore $G \subseteq U$. Further it follows for every $u \in G$ that

$$f_A(u)_k = \frac{e_k A u^T}{u A u^T} > u_k$$

which implies $f_A(G) \subseteq G$.

If we define a continuous function $V : \Delta \to \mathbb{R}$ by

$$V(u) = 1 - u_k \quad \text{for } u \in \Delta \, ,$$

then it follows that

$$V(f_A(u)) - V(u) = u_k - f_A(u)_k \leq 0 \quad \text{for all } u \in G \, .$$

Hence V is a Lyapunov function with respect to f_A on G.

Further it follows that

$$V(u) \geq 0 \text{ for all } u \in G \text{ and } (V(u) = 0 \Leftrightarrow u = e) \ ,$$

i.e., V is positive definite with respect to e_k. Finally we have

$$V(f_A(u)) - V(u) < 0 \quad \text{for all } u \in G \text{ with } u \neq e_k \ .$$

Thus all the assumptions of *Theorem 1.3* are satisfied and hence $\{e_k\}$ is asymptotically stable.

$\square$

3.7 A General Cooperative n-Person Goal-Cost-Game

3.7.1 The Game

In *Section 3.4.1* we have considered an n-person goal-cost-game that can be generalized as follows: Given n players who persue n goals which are given by an n-vector

$$b = (b_1, \ldots, b_n) \ .$$

In order to achieve these goals every player has to spend a certain amount of money, say $x_i \geq 0$ for the $i-th$ player. Every player P_i , $i = 1, \ldots, n$, can be assigned a goal value f_i which depends on the cost values of all players and can be described as a function $f_i : \mathbb{R}^n \rightarrow \mathbb{R}$ for $i = 1, \ldots, n$.
The requirement that all players reach their goal is assumed to be given by a system of inequalities of the form

$$f_i(x_1, \ldots, x_n) \geq b_i \quad \text{for } i = 1, \ldots, n \tag{3.77}$$

where

$$x_i \geq 0 \quad \text{for } i = 1, \ldots, n \ . \tag{3.78}$$

Every player is, of course, interested in minimizing his own cost subject to (3.77), (3.81). This, however, is in general simultaneously impossible. Therefore we assume in a first step that the players cooperate and minimize

$$s(x) = \sum_{i=1}^{n} x_i \tag{3.79}$$

subject to (3.77), (3.78).

Let us assume that $\hat{x} \in \mathbb{R}^n$ is a solution of this problem. If we then choose, for any $i \in \{1, \ldots, n\}$, some $x_i \geq 0$ such that

$$f_i(\hat{x}_1, \ldots, \hat{x}_{i-1}, x_i, \hat{x}_{i+1}, \ldots, \hat{x}_n) \geq b_i \, ,$$

it follows that

$$\sum_{j=1}^{n} \hat{x}_j \leq \sum_{\substack{j=1 \\ j \neq i}}^{n} \hat{x}_j + x_i$$

and therefore $\hat{x}_i \leq x_i$.

Thus every solution of (3.77), (3.78) which minimizes (3.79) is a Nash equilibrium, i.e., if the $i - th$ player declines from his choice of costs whereas all the others stick to it, he can at most do worse.

3.7.2 A Cooperative Treatment

Now we go one step further and define a cooperative $n-$person game in the following way: Let $\mathcal{N} = \{1, \ldots, n\}$. Then, for every non-empty subset S of $\mathcal{N}$, we define

$$f_S(x) = \sum_{i \in S} f_i(x) \, , \ x \in \mathbb{R}^n \, , \text{and } b_S = \sum_{i \in S} b_i$$

and consider the problem of minimizing

$$s(x) = \sum_{i \in \mathcal{N}} x_i \tag{3.79}$$

subject to (3.81) and

$$f_S(x_1, \ldots, x_n) \geq b_S \, . \tag{3.77}_S$$

Every non-empty subset S of $\mathcal{N}$ can be considered as a coalition in which the players join by adding their inequalities and minimizing (3.79). If we define, for every $S \subseteq \mathcal{N}$,

$$v(S) = \begin{cases} \inf\{ \sum_{i \in \mathcal{N}} x_i \mid x \in \mathbb{R}^n \text{ satisfies } (3.77)_S, (3.78)\}, & \text{if } S \text{ is non-empty,} \\ 0, & \text{if } S \text{ is empty,} \end{cases} \tag{3.80}$$

then $v : 2^{\mathcal{N}} \to \mathbb{R}_+$ is the payoff function of a cooperative $n-$person game. For the following we assume that, for every non-empty $S \subseteq \mathcal{N}$, there exists some $x \in \mathbb{R}^n$ with $(3.77)_S$, (3.81) and $\sum_{i \in \mathcal{N}} x_i = v(S)$.

The question now arises under which conditions the grand coalition $\mathcal{N}$ is stable which means that, if the players decide for a grand coalition there is no incentive for them to decline from this decision. This is certainly the case, if there is a division $\{x_1, \ldots, x_n\}$ of $v(\mathcal{N})$, i.e.,

$$x_i \geq 0 \quad \text{for } i = 1, \ldots, n \tag{3.81}$$

and

$$\sum_{i=1}^{n} x_i = v(\mathcal{N}) \tag{3.82}$$

such that

$$\sum_{i \in S} x_i \leq v(S) \text{ for all non-empty } S \subseteq \mathcal{N} . \tag{3.83}$$

If this condition is satisfied, the players have to spend at least as much as they have to spend in the grand coalition, if they choose another one.

3.7.3 Necessary and Sufficient Conditions for a Stable Grand Coalition

Theorem 3.7. *There exists a vector $x \in \mathbb{R}^n$ with (3.81), (3.82) and (3.83), if and only if for every collection $\mathcal{B} \subseteq 2^{\mathcal{N}}$ such that for every $S \in \mathcal{B}$ there exists a weight $\gamma_S \geq 0$ with*

$$\sum_{\substack{S \in \mathcal{B} \\ i \in S}} \gamma_S = 1 \quad \text{for all } i = 1, \ldots, n \tag{3.84}$$

it is true that

$$v(\mathcal{N}) \leq \sum_{S \in \mathcal{B}} \gamma_S v(S) . \tag{3.85}$$

Proof. 1) Let $x \in \mathbb{R}^n$ with $(3.81), (3.82), (3.83)$ be given.
Further let $\mathcal{B} \subset 2^{\mathcal{N}}$ be a collection such that for every $S \in \mathcal{B}$ there exists a weight $\gamma_S \geq 0$ with (3.84). Then it follows that

$$\sum_{S \in \mathcal{B}} \gamma_S v(S) \geq \sum_{S \in \mathcal{B}} \gamma_S \sum_{i \in \mathcal{B}} x_i = \sum_{i \in \mathcal{N}} \left(\sum_{\substack{S \in \mathcal{B} \\ i \in S}} \gamma_S \right) x_i$$

$$= \sum_{i \in \mathcal{N}} x_i = v(\mathcal{N}) .$$

Hence (3.85) is satisfied.

2) The proof of the sufficiency part is the same as that of *Theorem 3.5.*

□

As a generalization of *Lemma 3.1* we can prove

Lemma 3.4. *Let us assume that, for every* $i \in \{1, \ldots, n\}$, *for every finite sequence of vectors* $x^1, \ldots, x^m \in \mathbb{R}^n$ *and numbers* $\lambda_1 \geq 0, \ldots, \lambda_m \geq 0$ *it is true that*

$$f_i \left(\sum_{k=1}^{m} \lambda_k x^k \right) \geq \sum_{k=1}^{m} \lambda_k f_i(x^k) .$$

Further let us assume that

$$f_{\mathcal{N}}(x) \geq f_S(x)$$

for all non-empty $S \subseteq \mathcal{N}$ *and all* $x \in \mathbb{R}$ *with*

$$x_i \geq 0 \quad for \ i = 1, \ldots, n .$$

Then for every collection $\mathcal{B} \subseteq 2^{\mathcal{N}}$ *such that for every* $S \in \mathcal{B}$ *there exists a weight* $\gamma_S \geq 0$ *with (3.84) it is true that (3.85) is satisfied.*

Proof. Let $\mathcal{B} \subseteq 2^{\mathcal{N}}$ a collection as required.
Then it follows that

$$\sum_{S \in \mathcal{B}} \gamma_S b_S = \sum_{S \in \mathcal{B}} \sum_{i \in S} \gamma_S b_i = \sum_{i \in \mathcal{N}} \left(\sum_{\substack{S \in \mathcal{B} \\ i \in S}} \gamma_S \right) = b_{\mathcal{N}}$$

Now let, for every $S \in \mathcal{B}$,

$$v(S) = x_1(S) + \ldots + x_n(S)$$

where

$$x_i(S) \geq 0 , \quad i = 1, \ldots, n ,$$
$$f_S(x(S)) \geq b_S .$$

Then

$$\sum_{S \in \mathcal{B}} \gamma_S v(S) = \sum_{S \in \mathcal{B}} \gamma_S \sum_{j=1}^{n} x_j(S)$$

$$= \sum_{j=1}^{n} \left(\sum_{S \in \mathcal{B}} \gamma_S x_j(S) \right) = \sum_{j=1}^{n} \hat{x}_j$$

where

$$\hat{x}_j = \sum_{S \in \mathcal{B}} \gamma_S x_j(S) \quad for \ j = 1, \ldots, n .$$

Then it follows that

$$\hat{x}_j \geq 0 \quad \text{for } j = 1, \ldots, n$$

and

$$f_{\mathcal{N}}(\hat{x}) = f_{\mathcal{N}}\left(\sum_{S \in \mathcal{B}} \gamma_S x(S)\right) \geq \sum_{S \in \mathcal{B}} \gamma_S f_{\mathcal{N}}(x(S))$$

$$\geq \sum_{S \in \mathcal{B}} \gamma_S f_S(x(S)) \geq \sum_{S \in \mathcal{B}} \gamma_S b_S = b_{\mathcal{N}}$$

which implies

$$v(\mathcal{N}) \leq \sum_{j=1}^{n} \hat{x}_j = \sum_{S \in \mathcal{B}} \gamma_S v(S) \ ,$$

i.e., (3.85) is satisfied.

$\square$

3.8 A Cooperative Treatment of an n-Person Cost-Game

3.8.1 The Game and a First Cooperative Treatment

We consider n players P_i, $i = 1, \ldots, n$, $n \geq 2$, who play a game in which every player P_i has at his disposal a (non-empty) set $U_i \subseteq \mathbb{R}^{m_i}$ of strategies. They can, however, not necessarily choose their strategies independently of each other. If player P_i chooses $u_i \in U_i$ for $i = 1, \ldots, n$, then the n-tupel $(u_1, \ldots, u_n)$ is required to lie in a non-empty subset U of $\prod_{i=1}^{n} U_i$ which is assumed to be of the form

$$U = \bigcap_{i=1}^{n} V_i \quad \text{where} \quad V_i \subseteq \prod_{j=1}^{n} U_j \quad \text{for } i = 1, \ldots, n \ .$$

Further every player P_i is assigned a cost function $\varphi_i : \prod_{j=1}^{n} U_j \to \mathbb{R}_+$ which he wants to minimize on U. This, however, is in general impossible simultaneously. Therefore the players could minimize in a first step the function

$$\varphi(u) = \sum_{i=1}^{n} \varphi_i(u) \quad \text{for } u \in U .$$

If $\hat{u} \in U$ is such that

$$\varphi(\hat{u}) \leq \varphi(u) \quad \text{for all } u \in U ,$$

then it is easy to see that $\hat{u}$ is a so called Pareto optimum, i.e., if there is any $u \in U$ such that

$$\varphi_i(u) \leq \varphi_i(\hat{u}) \quad \text{for all } i = 1, \ldots, n ,$$

then it necessarily follows that

$$\varphi_i(u) = \varphi_i(\hat{u}) \quad \text{for all } i = 1, \ldots, n .$$

In *Section 3.4.1* we have considered the following special case:
Let $m_i = 1$ and $U_i = \mathbb{R}_+$ for every $i = 1, \ldots, n$. Further let $\varphi_i : \mathbb{R}_+^n \to \mathbb{R}_+$ be given by

$$\varphi_i(u_1, \ldots, u_n) = u_i \quad \text{for } i = 1, \ldots, n$$

and let V_i be given by

$$V_i = \{u \in \mathbb{R}_+^n \mid \sum_{j=1}^{n} c_{ij} u_j \geq b_i\} \quad \text{for } i = 1, \ldots, n .$$

Then

$$U = \{u \in \mathbb{R}_+^n \mid \sum_{j=1}^{n} c_{ij} u_j \geq b_i \quad \text{for all } i = 1, \ldots, n\} = \bigcap_{i=1}^{n} V_i .$$

In this case the minimization of

$$\varphi(u) = \sum_{i=1}^{n} u_i \quad \text{on} \quad U$$

also leads to a Nash equilibrium $\hat{u} \in U$ (see *Section 3.4.1*).

3.8.2 Transformation of the Game into a Cooperative Game

For every non-empty subset S of $\mathcal{N} = \{1, \ldots, n\}$ we choose a non-empty set $U_S \subseteq \bigcup_{i \in S} V_i$ (with a property to be specified later) and define

$$v(S) = \begin{cases} \inf\{\varphi(u) \mid u \in U_S\} & , \text{ if } S \text{ is non-empty}, \\ 0 & , \text{ if } S \text{ empty}. \end{cases}$$

Then $v : 2^N \to \mathbb{R}_+$ is the payoff function of a cooperative n-person game. In the above special case we define, for every non-empty $S \subseteq \mathcal{N}$,

$$c_j(S) = \sum_{i \in S} c_{ij} \quad \text{for } j = 1, \ldots, n \quad \text{and} \quad b(S) = \sum_{i \in S} b_i$$

and put

$$U_S = \{u \in \mathbb{R}_+^n \mid \sum_{j=1}^n c_j(S)u_j \geq b(S)\} .$$

Then it follows that

$$U_S \subseteq \bigcup_{i \in S} V_i .$$

Let us assume that U is non-empty. Then every U_S is non-empty and, since

$$\varphi(u) = \sum_{i=1}^n u_i \geq 0 \quad \text{for all} \quad u \in U_S ,$$

there exists some $\hat{u}_S \in U_S$ such that

$$\varphi(\hat{u}_S) = v(S) = \inf\{\varphi(u) \mid u \in U_S\} .$$

The question now arises under which conditions the grand coalition $\mathcal{N}$ is stable which means that, if the players decide for the grand coalition, there is no incentive for them to decline from this decision.

This is certainly the case, if there is a division $\{x_1, \ldots, x_n\}$ of $v(\mathcal{N})$, i.e.,

$$x_i \geq 0 \quad \text{for } i = 1, \ldots, n \tag{3.86}$$

and

$$\sum_{i=1}^n x_i = v(\mathcal{N}) \tag{3.87}$$

such that

$$\sum_{i \in S} x_i \leq v(S) \quad \text{for all non-empty } S \subseteq \mathcal{N} . \tag{3.88}$$

Every such division guarantees every player P_i a cost $x_i \leq v(\{i\})$ such that for every coalition $S \subseteq \mathcal{N}, S \neq \emptyset$ the joint cost $\sum_{i \in S} x_i$ is at most as high as $v(S)$.

3.8.3 Sufficient Conditions for a Stable Grand Coalition

We start with

Theorem 3.8. *There exists a vector $x \in \mathbb{R}^n$ with (3.86), (3.87), (3.88), if and only if the following condition is satisfied: If for every non-empty set $S \subseteq N$ there is a weight $\gamma_S \geq 0$ such that*

$$\sum_{\substack{S \in 2^N \setminus \{\emptyset\} \\ i \in S}} \gamma_S = 1 \quad \text{for all } i = 1, \ldots, n \, , \tag{3.89}$$

then it follows that

$$v(N) \leq \sum_{S \in 2^N \setminus \{\emptyset\}} \gamma_S v(S) \, . \tag{3.90}$$

The proof of the implication $((3.89) \Rightarrow (3.90)) \Rightarrow (3.86), (3.87), (3.88)$ is the same as that of *Theorem 3.5*. The proof of the implication $(3.86), (3.87), (3.88)$ $\Rightarrow ((3.89) \Rightarrow (3.90))$ has been given in *Section 3.4.5*. In order to apply this theorem to the cooperative n-person game defined in *Section 3.8.2* we make the following assumptions:

1) For every non-empty set $S \subseteq N$ let $\gamma_S \geq 0$ be a weight such that (3.89) is satisfied. Then for every $u_S \in U_S$ it follows that

$$\sum_{S \in 2^N \setminus \{\emptyset\}} \gamma_S u_S \in U_N \, .$$

2) For every non-empty set $S \subseteq N$ there is some $\hat{u}_S \in U_S$ with $\varphi(\hat{u}_S) = v(S)$.

3) For every $i \in N$ and every finite sequence $u^1, \ldots, u^m \in \mathbb{R}^{\sum\limits_{i=1}^{n} m_i}$ and numbers $\lambda_1 \geq 0, \ldots, \lambda_m \geq 0$ it is true that

$$\varphi_i \left(\sum_{k=1}^{m} \lambda_k u^k \right) \leq \sum_{k=1}^{m} \lambda_k \varphi_i(u^k) \, .$$

Then we can prove

Theorem 3.9. *If the assumptions 1), 2) and 3) are satisfied, then there is a vector $x \in \mathbb{R}^n$ with (3.86), (3.87), (3.88) which means that the grand coalition N is stable.*

Proof. Let, for every non-empty set $S \subseteq N$, a weight $\gamma_S \geq 0$ be given such that (3.89) holds true. Then it follows with $\mathcal{B} = 2^N \setminus \{\emptyset\}$ that

$$\sum_{S \in \mathcal{B}} \gamma_S v(S) = \sum_{S \in \mathcal{B}} \gamma_S \varphi(\hat{u}_S) \geq \varphi \left(\underbrace{\sum_{S \in \mathcal{B}} \gamma_S \hat{u}_S}_{\in U_N} \right) \geq v(N) \, , \text{ q.e.d. .}$$

$\square$

In the above special case assumption 2) is satisfied, if U is non-empty. Assumption 3) is obviously satisfied. Concerning assumption 1) we can prove

Lemma 3.5. *If*

$$c_{ii} > 0 \quad \text{for } i = 1, \ldots, n$$

and

$$c_{ij} \geq 0 \quad \text{for } i, j = 1, \ldots, n \ , \ i \neq j \ ,$$

(which implies that U is non-empty), then assumption 1) is also satisfied.

Proof. Let $\mathcal{B} = 2^N \setminus \{\emptyset\}$ and for every $S \in \mathcal{B}$ let there be given a weight $\gamma_S \geq 0$ such that

$$\sum_{\substack{S \in \mathcal{B} \\ i \in S}}^{n} \gamma_S = 1 \quad \text{for } i = 1, \ldots, n \ .$$

Then it follows that

$$\sum_{S \in \mathcal{B}} \gamma_S b(S) = \sum_{S \in \mathcal{B}} \gamma_S \sum_{i \in S} b_i = \sum_{i=1}^{n} \left(\sum_{\substack{S \in \mathcal{B} \\ i \in S}} \gamma_S \right) b_i = \sum_{i=1}^{n} b_i = b(N) \ .$$

Now let, for every $S \in \mathcal{B}$, be given some $u_S \in U_S$. Then it follows that

$$\sum_{j=1}^{n} c_j(S) \left(\sum_{S \in \mathcal{B}} \gamma_S u_S \right)_j = \sum_{S \in \mathcal{B}} \gamma_S \left(\sum_{j=1}^{n} c_j(S)(u_S)_j \right) \geq \sum_{S \in \mathcal{B}} \gamma_S b(S) = b(N) \ ,$$

hence

$$\sum_{j=1}^{n} c_j(S) \left(\sum_{S \in \mathcal{B}} \gamma_S u_S \right)_j \geq b(N)$$

which implies

$$\sum_{j=1}^{n} c_j(N) \left(\sum_{S \in \mathcal{B}} \gamma_S u_S \right)_j \geq b(N)$$

and hence $\sum_{S \in \mathcal{B}} \gamma_S u_S \in U_N$, since $\sum_{S \in \mathcal{B}} \gamma_S u_S \in \mathbb{R}_+^n$. This completes the proof.

$\square$

3.8.4 Further Cooperative Treatments

a) For every non-empty set $S \subseteq N$ we define

$$U_S = \bigcap_{i \in S} V_i$$

and put

$$v(S) = \inf\{\varphi(u) \mid u \in U_S\}$$

where again

$$\varphi(u) = \sum_{i \in N} \varphi_i(u) \quad \text{for } u \in \prod_{j=1}^{n} U_j .$$

Further we again put $v(\emptyset) = 0$.
From $U_S \subseteq V_i$ for all $i \in S$ it follows that

$$v(\{i\}) \leq v(S) \quad \text{for all } i \in S .$$

Let, for every non-empty set $S \subseteq N$,

$$\varepsilon_S = v(S) - \frac{1}{|S|} \sum_{i \in S} v(\{i\}) \quad (\geq 0) .$$

If we define

$$x_i = \frac{1}{n} \left(v(\{i\}) + \varepsilon_N \right) \quad \text{for } i = 1, \ldots, n ,$$

then it follows that

$$x_i \geq 0 \quad \text{for } i = 1, \ldots, n$$

and

$$\sum_{i=1}^{n} x_i = \frac{1}{n} \sum_{i=1}^{n} \left(v(\{i\}) + \varepsilon_N \right) = \frac{1}{n} \sum_{i=1}^{n} v(\{i\}) + \varepsilon_N = v(N) ,$$

hence (3.86) and (3.87) are satisfied.

Assumption:

$$\frac{\varepsilon_N}{n} \leq \frac{1}{|S|} \{\varepsilon_S + \sum_{i \in S} (\frac{1}{|S|} - \frac{1}{n}) v(\{i\})\} \quad \text{for all non-empty sets } S \subseteq N .$$

$$(3.91)$$

For every non-empty set $S \subseteq N$ it follows that

$$\sum_{i \in S} x_i = \sum_{i \in S} \frac{1}{n} \left(v(\{i\}) + \varepsilon_N \right)$$

$$\leq \frac{1}{n} \sum_{i \in S} v(\{i\}) + \varepsilon_S + \underbrace{\frac{1}{|S|} \sum_{i \in S} v(\{i\}) - \frac{1}{n} \sum_{i \in S} v(\{i\})}_{v(S)}$$

$$= v(S) ,$$

Result: If the condition (3.91) is satisfied, then the grand coalition is stable.

b) For every non-empty set $S \subseteq N$ we define

$$U_S = \bigcup_{i \in S} V_i .$$

Then it follows that, for every non-empty set $S \subseteq N$,

$$v(S) = \inf\{\varphi(u) \mid u \in U_S\} = \min_{i \in S} \inf\{\varphi(u) \mid u \in V_i = U_{\{i\}}\}$$

$$= \min_{i \in S} v(\{i\}) .$$

This implies

$$v(N) \leq v(S) \quad \text{for all non-empty sets } S \subseteq N .$$

If we define

$$x_i = \frac{1}{n} v(N) \quad \text{for } i = 1, \dots, n ,$$

then (3.86) and (3.87) are satisfied and for every non-empty set $S \subseteq N$ we obtain

$$\sum_{i \in S} x_i = \frac{|S|}{n} v(N) \leq v(S) , \quad \text{i.e. (3.88) is also satisfied.}$$

c) We again define, for every non-empty set $S \subseteq N$,

$$U_S = \bigcap_{i \in S} V_i .$$

Then we put

$$v(S) = \inf\{\varphi_S(u) \mid u \in U_S\} , \quad \text{if } S \subseteq N \quad \text{is non-empty} ,$$

where

$$\varphi_S(u) = \sum_{i \in S} \varphi_i(u) \ , \ u \in U_S \ .$$

Let us assume that, for every non-empty set $S \subseteq N$, there exists a $u(S) \in U_S$ such that $\varphi_S(u(S)) = v(S)$.
Since

$$U_S \subseteq V_i = U_{\{i\}} \quad \text{for every } i \in S \ ,$$

it follows that

$$v(\{i\}) \le \varphi_i(u(S)) \quad \text{for all } i \in S \ .$$

This implies

$$\sum_{i \in S} v(\{i\}) \le \varphi_S(u(S)) = v(S) \ .$$

If

$$\sum_{i \in N} v(\{i\}) = v(N) \ , \tag{3.92}$$

then $(v(\{1\}), \ldots, v(\{n\}))^T$ satisfies (3.86), (3.87), (3.88). Conversely, if this is the case, then (3.92) must hold true.

Result: Under the above assumption the condition (3.92) is necessary and sufficient for $(v(\{1\}), \ldots, v(\{n\}))^T$ to satisfies (3.86), (3.87), (3.88).
If (3.92) is satisfied, then it is easy to see that $v(\{1\}), \ldots, v(\{n\})$ is the only division of $v(N)$ which satisfies (3.88).

3.8.5 Pareto Optima as Cooperative Solutions of the Game

We consider the non-cooperative game in *Section 3.8.1* without assuming that the set $U \subseteq \prod_{i=1}^{n} U_i$ which restricts the strategies is of the form $U = \bigcap_{i=1}^{n} V_i$ with $V_i \subseteq \prod_{j=1}^{n} U_j$, for $i = 1, \ldots, n$.
We have seen that the minimization of

$$\varphi(u) = \sum_{i=1}^{n} \varphi_i(u) \quad \text{for } u \in U$$

leads to a Pareto optimum $\hat{u} \in U$ which has the property that for every $u \in U$ such that

$$\varphi_i(u) \le \varphi_i(\hat{u}) \quad \text{for all } i = 1, \ldots, n \tag{3.93}$$

it follows that

$$\varphi_i(u) = \varphi_i(\hat{u}) \quad \text{for all } i = 1, \ldots, n \ . \tag{3.94}$$

By contraposition this is equivalent to the following statement: If for $u \in U$ there exists $i_0 \in \{1, \ldots, n\}$ with $\varphi_{i_0}(u) < \varphi_{i_0}(\hat{u})$, then there is some $i_1 \in \{1, \ldots, n\}$ with $\varphi_{i_1}(u) > \varphi_{i_0}(\hat{u})$. This means that there is no $u \in U$ for which a player improves his cost value without the cost value of at least one other player deteriorating.

In this sense a Pareto optimum can be considered as a cooperative solution of the game. A sufficient condition for some $\hat{u} \in U$ to be a Pareto optimum is given in

Theorem 3.10. *Let $\hat{u} \in U$ be such that there exist numbers $y_i > 0$ for $i = 1, \ldots, n$ with*

$$\sum_{i=1}^{n} y_i \varphi_i(\hat{u}) \le \sum_{i=1}^{n} y_i \varphi_i(u) \quad \text{for all } u \in U \ . \tag{3.95}$$

Then $\hat{u}$ is a Pareto optimum.

Proof. Let $u \in U$ be given such that (3.93) holds true. Then it follows that

$$\sum_{i=1}^{n} y_i (\varphi_i(\hat{u}) - \varphi_i(u)) \ge 0 \ .$$

From (3.95) it follows that

$$\sum_{i=1}^{n} y_i (\varphi_i(\hat{u}) - \varphi_i(u)) \le 0 \ ,$$

hence

$$\sum_{i=1}^{n} y_i (\varphi_i(\hat{u}) - \varphi_i(u)) = 0 \ ,$$

which implies, together with (3.93), that (3.94) must hold true.

$$\square$$

Conversely we have the

Theorem 3.11. *Let the set $\Phi(U) + \mathbb{R}_+^n$ with*

$$\Phi(u) = \begin{pmatrix} \varphi_1(u) \\ \vdots \\ \varphi_n(u) \end{pmatrix} , \quad u \in U \ ,$$

be convex and have a non-empty interior $int(\Phi(U) + \mathbb{R}_+^n)$.

Assertion: If $\hat{u} \in U$ is a Pareto optimum, then there is a vector $y \in \mathbb{R}_+^n$ with $y \neq \Theta_n$ such that (3.95) holds true.

Proof. $\hat{u} \in U$ being a Pareto optimum implies that

$$\{\Phi(\hat{u})\} = \left(\Phi(\hat{u}) - \mathbb{R}_+^n\right) \cap \left(\Phi(U) + \mathbb{R}_+^n\right) \ .$$

Since $\Phi(\hat{u}) \notin int\left(\Phi(U) + \mathbb{R}_+^n\right)$, it follows that

$$\left(\Phi(\hat{u}) - \mathbb{R}_+^n\right) \cap \ int\left(\Phi(U) + \mathbb{R}_+^n\right) = \emptyset \ .$$

Since both sets are convex, there exists a vector $y \in \mathbb{R}^n$ with $y \neq \Theta_n$ and some $\alpha \in \mathbb{R}$ such that

$$y^T\left(\Phi(\hat{u}) - z_1\right) \leq \alpha \leq y^T\left(\Phi(u) + z_2\right) \text{ for all } z_1, z_2 \in \mathbb{R}_+^n \text{ and all } u \in U \ .$$

This implies $y \in \mathbb{R}_+^n$ and

$$y^T\Phi(\hat{u}) \leq y^T\Phi(u) \quad \text{for all } u \in U \ . \tag{3.96}$$

This completes the proof.

$\square$

The notion of Pareto optimum can be weakened in the following way:

An element $\hat{u} \in U$ is called weak Pareto optimum, if there is no $u \in U$ such that

$$\varphi_i(u) < \varphi_i(\hat{u}) \quad \text{for all } i = 1, \dots, n \ . \tag{3.97}$$

Obviously a Pareto optimum is also a weak Pareto optimum. The converse is false, in general.

Theorem 3.12. *Let $\hat{u} \in U$ be such that there exists a vector $y \in \mathbb{R}_+^n$ with $y \neq \Theta_n$ such (3.96) holds true. Then $\hat{u}$ is a weak Pareto optimum.*

Proof. Let $u \in U$ be such that (3.97) is satisfied. Then it follows that

$$y^T\Phi(u) < y^T\Phi(\hat{u})$$

which contradicts (3.96). This completes the proof.

$\square$

Conversely we have the

Theorem 3.13. *Let the set $\Phi(U) + \mathbb{R}_+^n$ be convex.*
Assertion: If $\hat{u} \in U$ is a weak Pareto optimum, than there is a vector $y \in \mathbb{R}_+^n$ with $y \neq \Theta_n$ such that (3.96) holds true.

Proof. From $\hat{u} \in U$ being a weak Pareto optimum it follows that

$$\left(\Phi(\hat{u}) - \overset{\circ}{\mathbb{R}}_+^n \right) \cap \left(\Phi(U) + \mathbb{R}_+^n \right) = \emptyset .$$

The rest of the proof is the same as that of *Theorem 3.11.*

□

A

Appendix

A.1 The Core of a Cooperative n-Person Game

In *Section 3.4.3* and *3.4.4* we have investigated the core of a special cooperative n-person game and have given necessary and sufficient conditions for its
non-emptiness. Here we consider a general cooperative n-person game repre-
sented by a function $v : 2^N \to \mathbb{R}$ where $N = \{1, \ldots, n\}$, $n \geq 2$, with $v(\emptyset) = 0$.
The core $C(v)$ of such a game is given by all vectors $x \in \mathbb{R}^n$ with

$$\sum_{i \in N} x_i = v(N) \tag{A.1}$$

and

$$\sum_{i \in S} x_i \geq v(S) \quad \text{for all non-empty } S \subseteq N . \tag{A.2}$$

In [3] there is given a necessary and sufficient condition for the core $C(v)$ to be
non-empty (see *Chapter II, Theorems 8.3 and 8.4*) which can be formulated
as follows:

Theorem A.1: There exists a vector $x \in \mathbb{R}^n$ with (A.1) and (A.2), if and
only if for every collection $\mathcal{B} \subseteq 2^N \setminus \{\emptyset\}$ and every set of weights $\gamma_S \geq 0$,
$S \in \mathcal{B}$, with

$$\sum_{\substack{S \in \mathcal{B} \\ i \in S}}^{n} \gamma_S = 1 \quad \text{for } i = 1, \ldots, n \tag{A.3}$$

it follows that

$$\sum_{S \in \mathcal{B}} \gamma_S v(S) \leq v(N) . \tag{A.4}$$

Proof.

1) Let $x \in \mathbb{R}^n$ be given such that (A.1) and (A.2) are satisfied. Let further $\mathcal{B} \subseteq 2^N \setminus \{\emptyset\}$ and $\{\gamma_S \geq 0 \mid S \in \mathcal{B}\}$ be given such that (A.3) holds true. Then it follows that

$$v(N) = \sum_{i \in N} x_i = \sum_{i \in N} \left(\sum_{\substack{S \in \mathcal{B} \\ i \in S}} \gamma_S \right) x_i = \sum_{S \in \mathcal{B}} \gamma_S \sum_{i \in S} x_i \geq \sum_{S \in \mathcal{B}} \gamma_S v(S) ,$$

   hence (A.4) holds true.

2) In order to show the sufficiency of the implication (A.3) $\Rightarrow$ (A.4) for the existence of some $x \in \mathbb{R}^n$ with (A.1), (A.2) we consider the problem of minimizing $\sum\limits_{i \in N} x_i$ subject to (A.2).

   The dual to this problem consists of maximizing

$$\sum_{S \in 2^N \setminus \{\emptyset\}} \gamma_S v(S)$$

subject to

$$\gamma_S \geq 0 \quad \text{for all } S \in 2^N \setminus \{\emptyset\} \tag{A.5}$$

and

$$\sum_{\substack{S \in 2^N \setminus \{\emptyset\} \\ i \in S}} \gamma_S = 1 \quad \text{for } i = 1, \ldots, n . \tag{A.6}$$

Obviously there exist vectors $x \in \mathbb{R}^n$ such that (A.2) is satisfied and for every such vector we have that

$$\sum_{i \in N} x_i \geq v(N) .$$

Therefore there exists a solution of the above problem and it is

$$\min\{\sum_{i \in N} x_i \mid x \in \mathbb{R}^n \quad \text{satisfies (A.2)}\} \geq v(N) .$$

By a duality theorem of linear programming the dual problem also has a solution and it is

$$\max\{\sum_{S \in 2^N \setminus \{\emptyset\}} \gamma_S v(S) \mid (\gamma_S)_{S \in 2^N \setminus \{\emptyset\}} \text{ satisfies (A.5) , (A.6)}\}$$

$$= \min\{\sum_{i \in N} x_i \mid x \in \mathbb{R}^n \text{ satisfies (A.2)}\} \geq v(N) .$$

From the implication (A.3) $\Rightarrow$ (A.4) it follows that

$$\max\{ \sum_{S \in 2^N \setminus \{\emptyset\}} \gamma_S v(S) \mid (\gamma_S)_{S \in 2^N \setminus \{\emptyset\}} \text{ satisfies } (A.5) \text{ and } (A.6)\} \le v(N)$$

which then implies

$$\min\{\sum_{i \in N} x_i \mid x \in \mathbb{R}^n \text{ satisfies } (A.2)\} = v(N)$$

and completes the proof.

$\square$

The dual problem can be used in order to find a vector $x \in \mathbb{R}^n$ with (A.1), (A.2), if such one exists. Let us demonstrate this for the case $n = 3$, $v(\{1\}) = v(\{2\}) = v(\{3\}) = 0$, $v(\{1,2\}) > 0$, $v(\{1,3\}) > 0$, $v(\{2,3\}) > 0$ and $v(N) > 0$, $N = \{1,2,3\}$. In this case the core $C(v)$ consists of all vectors $x \in \mathbb{R}^3$ such that

$$
\begin{aligned}
x_1 + x_2 + x_3 &= v(N) \ , \\
x_1 + x_2 \phantom{{}+ x_3} &\ge v(\{1,2\}) \ , \\
x_1 \phantom{{}+ x_2} + x_3 &\ge v(\{1,3\}) \ , \\
x_2 + x_3 &\ge v(\{2,3\}) \ ,
\end{aligned}
\tag{A.7}
$$

$$x_1 \ge 0 \ , \ x_2 \ge 0 \ , \ x_3 \ge 0 \ .$$

The dual problem consists of maximizing

$$\sum = \gamma_{\{1,2\}} v(\{1,2\}) + \gamma_{\{1,3\}} v(\{1,3\}) + \gamma_{\{2,3\}} v(\{2,3\}) + \gamma_N v(N)$$

subject to

$$
\begin{aligned}
\gamma_{\{1,2\}} + \gamma_{\{1,3\}} \phantom{{}+ \gamma_{\{2,3\}}} + \gamma_N &\le 1 \ , \\
\gamma_{\{1,2\}} \phantom{{}+ \gamma_{\{1,3\}}} + \gamma_{\{2,3\}} + \gamma_N &\le 1 \ , \\
\gamma_{\{1,3\}} + \gamma_{\{2,3\}} + \gamma_N &\le 1 \ ,
\end{aligned}
$$

$$\gamma_{\{1,2\}} \ge 0 \ , \ \gamma_{\{1,3\}} \ge 0 \ , \ \gamma_{\{2,3\}} \ge 0 \ , \ \gamma_N \ge 0 \ .$$

In order to solve the dual problem with the aid of the simplex method we introduce slack variables

$$z_1 \ge 0 \ , \ z_2 \ge 0 \ , \ z_3 \ge 0$$

and rewrite the constraints in the form

$$
\begin{aligned}
\gamma_{\{1,2\}} + \gamma_{\{1,3\}} \phantom{{}+ \gamma_{\{2,3\}}} + \gamma_N + z_1 &= 1 \\
\gamma_{\{1,2\}} \phantom{{}+ \gamma_{\{1,3\}}} + \gamma_{\{2,3\}} + \gamma_N + z_2 &= 1 \\
\gamma_{\{1,3\}} + \gamma_{\{2,3\}} + \gamma_N + z_3 &= 1
\end{aligned}
$$

The starting tableau for the simplex method then reads

		$-\gamma_{\{1,2\}}$	$-\gamma_{\{1,3\}}$	$-\gamma_{\{2,3\}}$	$-\gamma_N$
z_1	1	1	1	0	$\boxed{1}$
z_2	1	1	0	1	1
z_3	1	0	1	1	1
$\sum$	0	$-v(\{1,2\})$	$-v(\{1,3\})$	$-v(\{2,3\})$	$-v(N)$

If we exchange γ_N with z_1, we obtain the tableau

		$-\gamma_{\{1,2\}}$	$-\gamma_{\{1,3\}}$	$-\gamma_{\{2,3\}}$	$-z_1$
γ_N	1	1	1	0	1
z_2	0	0	-1	$\boxed{1}$	-1
z_3	0	-1	0	1	-1
$\sum$	$v(N)$	$v(N)-v(\{1,2\})$	$v(N)-v(\{1,3\})$	$-v(\{2,3\})$	$v(N)$

If we exchange $\gamma_{\{2,3\}}$ with z_2, we obtain the tableau

		$-\gamma_{\{1,2\}}$	$-\gamma_{\{1,3\}}$	$-z_2$	$-z_1$
γ_N	1	1	1	0	1
$\gamma_{\{2,3\}}$	0	0	-1	1	-1
z_3	0	-1	$\boxed{1}$	-1	0
$\sum$	$v(N)$	$v(N)$ $-v(\{1,2\})$	$v(N)-v(\{1,3\})$ $-v(\{2,3\})$	$v(\{2,3\})$	$v(N)$ $-v(\{2,3\})$

If we exchange $\gamma_{\{1,3\}}$ with z_3, we obtain the tableau

		$-\gamma_{\{1,2\}}$	$-z_3$	$-z_2$	$-z_1$
γ_N	1	2	-1	1	1
$\gamma_{\{2,3\}}$	0	-1	1	0	-1
$\gamma_{\{1,3\}}$	0	-1	1	-1	0
$\sum$	$v(N)$	$2v(N)-v(\{1,2\})-$ $v(\{1,3\})-v(\{2,3\})$	$v(\{1,3\})+$ $v(\{2,3\})-v(N)$	$v(N)-$ $v(\{1,3\})$	$v(N)-$ $v(\{2,3\})$

If we assume that

$$2v(N) - v(\{1,2\}) - v(\{1,3\}) - v(\{2,3\}) \geq 0 \ ,$$

$$v(\{1,3\}) + v(\{2,3\}) - v(N) \geq 0 \ ,$$
$$v(N) - v(\{1,3\}) \geq 0 \ ,$$
$$v(N) - v(\{2,3\}) \geq 0 \ ,$$

then $\gamma_N = 1$, $\gamma_{\{1,2\}} = \gamma_{\{1,3\}} = \gamma_{\{2,3\}} = 0$ is a solution of the dual problem and

$$x_1 = v(N) - v(\{2,3\}) \ ,$$
$$x_2 = v(N) - v(\{1,3\}) \ ,$$
$$x_3 = v(\{1,3\}) + v(\{2,3\}) - v(N)$$

satisfy (A.7).

A.2 The Core of a Linear Production Game

We consider a linear production game with n players. Each player has at his disposal a vector $b^i = (b_1^i, b_2^i, \ldots, b_m^i)$, $i = 1, \ldots, n$, of resources $b_k^i > 0$, $k = 1, \ldots, m$, which he can use to produce goods that can be sold at a given market price. This section is taken from [24].

We assume that a unit of the j-th good $(j = 1, \ldots, p)$ requires $a_{kj} \geq 0$ units of the k-th resource $(k = 1, \ldots, m)$ and can be sold at a price $c_j > 0$.
Let $S \subseteq N = \{1, \ldots, n\}$, $S \neq \emptyset$, be a coalition. This coalition then has a total of

$$b_k(S) = \sum_{i \in S} b_k^i$$

units of the k-th resource. Using all of their resources, the members of S can produce vectors $(x_1, x_2, \ldots, x_p)$ of goods which satisfy

$$\sum_{j=1}^p a_{kj} x_j \leq b_k(S) \quad \text{for } k = 1, \ldots, m , \tag{A.8}$$

$$x_j \geq 0 \quad \text{for } j = 1, \ldots, p .$$

Under these conditions they want to maximize their profit

$$\sum_{j=1}^p c_j x_j .$$

If we define

$$v(S) = \max\{\sum_{j=1}^p c_j x_j \mid x \in \mathbb{R}^p \text{ satisfies } (A.8)\} ,$$

if S is non-empty (in which case the problem of maximizing

$$\sum_{j=1}^p c_j x_j$$

subject to (A.8) has a solution, if for every $k \in \{1, \ldots, m\}$ there is at least one $j \in \{1, \ldots, p\}$ such that $a_{kj} > 0$), and $v(\emptyset) = 0$, then $v : 2^N \to \mathbb{R}_+$ is the characteristic function of a cooperative n-person game.

For this game we can prove

Theorem A.2: The core of this game is non-empty.

Proof. We make use of *Theorem A.1* and consider an arbitrary collection $\mathcal{B} \subseteq 2^N \setminus \{\emptyset\}$ and weights $\gamma_S \geq 0$, $S \in \mathcal{B}$ with (A.3).

For every $k = 1, \ldots, m$ we then have

$$
\begin{aligned}
\sum_{S \in \mathcal{B}} \gamma_S b_k(S) &= \sum_{S \in \mathcal{B}} \sum_{i \in S} \gamma_S b_k^i \\
&= \sum_{i \in N} \{ \sum_{\substack{S \in \mathcal{B} \\ i \in S}} \gamma_S \} b_k^i \\
&= \sum_{i \in N} b_k^i \\
&= b_k(N) \; .
\end{aligned}
\tag{A.9}
$$

Now let $x_1(S), \ldots, x_p(S)$ be such that (A.8) is satisfied and

$$
v(S) = \sum_{j=1}^{p} c_j x_j(S) \; .
$$

Then

$$
\begin{aligned}
\sum_{S \in \mathcal{B}} \gamma_S v(S) &= \sum_{S \in \mathcal{B}} \{ \gamma_S \sum_{j=1}^{p} c_j x_j(S) \} \\
&= \sum_{j=1}^{p} c_j \{ \sum_{S \in \mathcal{B}} \gamma_S x_j(S) \} \\
&= \sum_{j=1}^{p} c_j \hat{x}_j
\end{aligned}
$$

where

$$
\hat{x}_j = \sum_{S \in \mathcal{B}} \gamma_S x_j(S) \quad \text{for } j = 1, \ldots, p \; .
$$

Now it follows that, for every $k = 1, \ldots, m$,

$$
\sum_{j=1}^{n} a_{kj} \left(\sum_{S \in \mathcal{B}} \gamma_S x_j(S) \right)
$$

$$
= \sum_{S \in \mathcal{B}} \gamma_S \left(\sum_{j=1}^{p} a_{kj} x_j(S) \right) \tag{A.10}
$$

$$
\leq \sum_{S \in \mathcal{B}} \gamma_S b_k(S)
$$

which implies

$$
\sum_{j=1}^{p} a_{kj} \hat{x}_j \leq b_k(N) \quad \text{for all } k = 1, \ldots, m ,
$$

Since

$$
\hat{x}_j \geq 0 \quad \text{for } j = 1, \ldots, p ,
$$

it follows that

$$
\sum_{j=1}^{p} c_j \hat{x}_j \leq v(N)
$$

which implies

$$
\sum_{S \in \mathcal{B}} \gamma_S v(S) = \sum_{j=1}^{p} c_j \hat{x}_j \leq v(N) .
$$

Hence (A.4) is satisfied. By *Theorem A.1* therefore the core is non-empty which completes the proof.

$\square$

In order to find points in the core we consider for every $S \subseteq N$, $S \neq \emptyset$, the dual of the problem of maximizing $\sum_{j=1}^{p} c_j x_j$ subject to (A.8) which consists of minimizing

$$
\sum_{k=1}^{m} b_k(S) y_k
$$

subject to

$$
\sum_{k=1}^{m} a_{kj} y_k \geq c_j \quad \text{for } j = 1, \ldots, p ,
$$

$$
y_1 \geq 0, \ldots, y_m \geq 0 .
$$

Let $y_1(S), \ldots, y_m(S)$ be a solution of this problem (which exists). Then it follows for $S = N$

$$v(N) = \sum_{k=1}^{m} b_k(N) y_k(N)$$

and, for every $S \subseteq N$, $S \neq \emptyset$, N.

$$v(S) \leq \sum_{k=1}^{m} b_k(S) y_k(N) \ .$$

Now let us define, for every $i = 1, \ldots, n$,

$$u_i = \sum_{k=1}^{m} b_k^i y_k(N) \ .$$

Then it follows, for every $S \subseteq N$, $S \neq \emptyset$,

$$\sum_{i \in S} u_i = \sum_{i \in S} \sum_{k=1}^{m} b_k^i y_k(N)$$

$$= \sum_{k=1}^{m} (\sum_{i \in S} b_k^i) y_k(N)$$

$$= \sum_{k=1}^{m} b_k(S) y_k(N)$$

and therefore

$$\sum_{i \in N} u_i = v(N)$$

and

$$\sum_{i \in S} u_i \geq v(S) \quad \text{for every } S \text{ non-empty } \subseteq N \ .$$

Thus $(u_1, \ldots, u_n)$ is a point in the core.

A.3 Weak Pareto Optima: Necessary and Sufficient Conditions

In *Section 3.8.5* we have given necessary and sufficient conditions for weak Pareto optima of non-cooperative n-person games. Here we will derive further conditions. We consider again a non-cooperative n-person game with n payoff functions

$$\varphi_i : \mathbb{R}^M \to \mathbb{R} \, , \; i = 1, \ldots, n \, ,$$

which are to be minimized on a non-empty set $U \subseteq \mathbb{R}^M$.
Then a weak Pareto optimum is a vector $\hat{u} \in U$ such that there is no vector $u \in U$ with

$$\varphi_i(u) < \varphi_i(\hat{u}) \quad \text{for all } i = 1, \ldots, n \, .$$

Now we assume that $\varphi_i \in C^1(\mathbb{R}^M)$ for $i = 1, \ldots, n$ and that U is convex. Then we can prove the following

Theorem A.3: If $\hat{u} \in U$ is a weak Pareto optimum, then there is no $u \in U$ such that

$$\varphi_i'(\hat{u})^T (u - \hat{u}) < 0 \quad \text{for all } i = 1, \ldots, n \tag{A.11}$$

Proof. We assume that there exists some $u \in U$ such that (A.9) is satisfied. If we define, for every $\lambda \in (0, 1]$, $u_\lambda = \hat{u} + \lambda(u - \hat{u})$, then it follows that

$$u_\lambda \in U \quad \text{for all } \lambda \in (0, 1] \, .$$

Further we have

$$\lim_{\lambda \to 0} \frac{1}{\lambda}(\varphi_i(u_\lambda) - \varphi_i(\hat{u})) = \varphi_i'(\hat{u})^T (u - \hat{u}) < 0$$

for all $i = 1, \ldots, n$.

Therefore for every $i = 1, \ldots, n$ there exists some $\lambda_i \in (0, 1]$ such that

$$\frac{1}{\lambda}(\varphi_i(u_\lambda) - \varphi_i(\hat{u})) < 0 \quad \text{for all } \lambda \in (0, \lambda_i] \, .$$

If we put $\lambda_{i_0} = \min_{i=1,\ldots,n} \lambda_i$, then it follows that

$$\varphi_i(u_\lambda) < \varphi_i(\hat{u}) \quad \text{for all } \lambda \in (0, \lambda_{i_0}] \quad \text{and all } i = 1, \ldots, n$$

which contradicts the assumption that $\hat{u}$ is a weak Pareto optimum.

If all payoff functions φ_i are convex and for some given $\hat{u} \in U$ there is no $u \in U$ with (A.9), then $\hat{u}$ is a weak Pareto optimum. This follows immediately from

$$\varphi_i(u) - \varphi_i(\hat{u}) \le \varphi_i'(\hat{u})^T(u - \hat{u}) \text{ for all } u \in U \text{ and all } i = 1, \dots, n .$$

The condition (A.9) not being satisfied, is equivalent to

$$(\Phi'(\hat{u})(\hat{u}) - \overset{\circ}{\mathcal{R}}{}_+^n) \cap \Phi'(\hat{u})(U) = \emptyset \tag{A.12}$$

where

$$\Phi'(\hat{u})(u) = \begin{pmatrix} \varphi_1'(\hat{u})^T u \\ \vdots \\ \varphi_n'(\hat{u})^T u \end{pmatrix} \quad \text{for all } u \in U .$$

Since $\Phi'(\hat{u})(U)$ is convex, (A.10) implies the existence of some $y \in \mathbb{R}_+^n$ with $y \ne \Theta_n$ such that

$$y^T \Phi'(\hat{u})(\hat{u}) \le y^T \Phi'(\hat{u})(u) \text{ for all } u \in U \tag{A.13}$$

where

$$y^T \Phi'(\hat{u})(u) = \left(\sum_{i=1}^n y_i \varphi_i'(\hat{u}) \right)^T u , \; u \in U .$$

Conversely, the existence of some $y \in \mathbb{R}_+^n$ with $y \ne \Theta_n$ with (A.11) implies that there is no $u \in U$ with (A.9).

$\square$

A.4 Duality

We consider the same non-cooperative n-person game as in *Section A.3* and define

$$\Phi(u) = \begin{pmatrix} \varphi_1(u) \\ \vdots \\ \varphi_n(u) \end{pmatrix} \quad \text{for } u \in U .$$

Then $\hat{u} \in U$ is a Pareto optimum, if and only if

$$(\Phi(\hat{u}) - \mathbb{R}_+^n) \cap \Phi(U) = \{\Phi(\hat{u})\} . \tag{A.14}$$

Let us define

$$D = \mathbb{R}^n \setminus (\Phi(U) + (\mathbb{R}_+^n \setminus \{\emptyset\})) .$$

Then it follows that

$$(\Phi(U) + (\mathbb{R}_+^n \setminus \{\emptyset_n\})) \cap D = \emptyset = \text{ empty set } .$$

Now let $\hat{u} \in U$ be such that $\Phi(\hat{u}) \in D$. Then it follows that

$$(\Phi(\hat{u}) + (\mathbb{R}_+^n \setminus \{\Theta_n\})) \cap D = \emptyset$$

which is equivalent to

$$(\Phi(\hat{u}) + \mathbb{R}_+^n) \cap D = \{\Phi(\hat{u})\} \tag{A.15}$$

and in turn to the implication

$$(\Phi(\hat{u}) \leq d \quad \text{for some } d \in D) \Rightarrow \Phi(\hat{u}) = d . \tag{A.16}$$

Conversely, the condition (A.13) implies $\Phi(\hat{u}) \in D$, $\hat{u} \in U$, which is equivalent to

$$\Phi(\hat{u}) \notin \Phi(U) + (\mathbb{R}_+^n \setminus \{\Theta_n\})$$

and in turn equivalent to

$$\Phi(\hat{u}) - (\mathbb{R}_+^n \setminus \{\Theta_n\}) \cap \Phi(U) = \emptyset$$

which is equivalent to (A.12).

Result: A vector $\hat{u} \in U$ is a Pareto optimum, if and only if the implication (A.14) holds true.

An element $\hat{u} \in U$ is a weak Pareto optimum, if and only if

$$(\Phi(\hat{u}) - \overset{\circ}{\mathbb{R}}{}_+^n) \cap \Phi(U) = \emptyset .$$

Let us define the set

$$D = \mathbb{R}^n \setminus (\Phi(U) + \overset{\circ}{\mathbb{R}}{}_+^n) .$$

Then it follows that

$$(\Phi(U) + \overset{\circ}{\mathbb{R}}{}_+^n) \cap D = \emptyset .$$

Now let $\hat{u} \in U$ be such that $\Phi(\hat{u}) \in D$. Then it follows that

$$(\Phi(\hat{u}) + \overset{\circ}{\mathbb{R}}{}_+^n) \cap D = \emptyset \tag{A.17}$$

which is equivalent to the non-existence of a $d \in D$ such that $\varphi_i(\hat{u}) < d_i$ for all $i = 1, \ldots, n$. $\Phi(\hat{u}) \in D$ for some $\hat{u} \in U$ is equivalent with

$$\Phi(\hat{u}) \notin \Phi(U) + \overset{\circ}{\mathbb{R}}{}_+^n$$

and in turn equivalent with

$$(\Phi(\hat{u}) - \overset{\circ}{\mathbb{R}}{}_+^n) \cap \Phi(U) = \emptyset .$$

Result: If $\hat{u} \in U$ is a weak Pareto optimum, then (A.15) holds true, i.e., there exists no $d \in D$ such that (A.16) holds true.

B

Bibliographical Remarks

The definition of time discrete autonomous dynamical systems given in *Section 1.1.1* is a special case of the abstract definition of dynamical systems which goes back to G. D. Birkhoff [2] and can also be found in the essay "What is a Dynamical System?" by G. R. Sell in [8]. In this essay also the time discrete case is discussed.

The localization of limit sets with the aid of Lyapunov functions in *Section 1.1.2* has been first represented by J. P. La Salle in [8] and can also be found in [13] and [20].

The stability results given in *Section 1.1.3* generalize a standard result on the stability of fixed points to be found in [8], [13], [18], [20] (see *Theorem 1.3*).

Theorem 1.5 in *Section 1.1.4* coincides with *Satz 5.4* in [18], if we choose as normed linear space the n-space $\mathbb{R}^n$ equipped with any norm.

The results on linear systems following the Corollary of *Theorem 1.6* in *Section 1.1.5* and also those for non-autonomous systems in *Section 1.2.3* are based on *Section 4.2* in [18].

The stability results in *Section 1.2.2* generalize those of *Section 1.1.3* and are taken from [15]. Lyapunov's method has also been applied to non-autonomous systems in [1] and [19] in order to investigate stability of fixed points. Instead of one Lyapunov function a sequence of such is used.

Theorem 2.2 about null-controllability of linear systems has been taken from [14]. *Proposition 2.3* concerning Kalman's condition can be found in [20] as well as *Propositions 2.1* and *2.2*. *Theorem 2.5* which generalizes *Theorem 2.1* to non-autonomous systems is taken from [15] and further generalized in *Section 2.2.2*. The results on stabilization of controlled systems in *Section 2.2.3* are taken from [15].

The dynamical games which are introduced in *Section 3.1* have also been investigated in [13], however, in a different way.

The cooperative treatment of the emission reduction model in *Sections 3.4.2* and *3.4.3* can also be found in [25]. The non-cooperative game in *Section 3.4.1* and its cooperative treatment in *Section 3.4.5* is also contained in [16]. The dynamical method for finding an evolutionary stable state in an evolution matrix game which is described in *Section 3.6.2* has been adopted from [21].

References

1. R.P. Agarwal: Difference Equations and Inequalities.
   Marcel Dekker, Inc.: New York, Basel, Hongkong 1992.
2. G. D. Birkhoff: Dynamical Systems.
   Amer. Math. Soc. Colloq. Publ. Providence 1927.
3. Th. Driessen: Cooperative Games, Solutions and Applications.
   Kluwer Academic Publishers, Dordrecht - Boston - London 1988.
4. U. Faigle, W. Kern, and G. Still: Algorithmic Principles of
   Mathematical Programming.
   Kluwer Academic Publishers: Dordrecht - Boston - London 2002.
5. F. R. Gantmacher: Application of the Theory of Matrices.
   Interscience Publishers: New York - London - Sydney 1959.
6. G.H. Golub, C.F. Van Loan: Matrix Computations.
   The Johns Hopkins University Press 3rd. ed. 1996.
7. Ch. Großmann, J. Terno.: Numerik der Optimierung.
   B.G. Teubner Stuttgart 1997.
8. J. Hale (editor): Studies in Ordinary Differential Equations.
   Vol. 14, published by the Mathematical Association of America, 1977.
9. J. Hülsmann, W. Gamerith, U. Leopold-Wildburger, W. Steindl: Einführung
   in die Wirtschaftsmathematik.
   3. Auflage Springer Verlag: Berlin - Heidelberg - New York 2002.
10. J. Jahn: Introduction to the theory of nonlinear optimization.
    (2-nd. edition). Academic Press, New York, 1994.
11. W. Krabs: Einführung in die lineare und nichtlineare
    Optimierung für Ingenieure.
    Verlag B. G. Teubner, Stuttgart 1983.
12. W. Krabs: Mathematische Modellierung.
    Verlag B. G. Teubner: Stuttgart 1997.
13. W. Krabs: Dynamische Systeme: Steuerbarkeit und chaotisches Verhalten.
    Verlag B. G. Teubner: Stuttgart und Leipzig 1998.
14. W. Krabs: On Local Controllability of Time-Discrete Dynamical Systems into
    Steady States.
    Journal of Difference Equations and Applications 8, no.1, 1 - 11 (2002).
15. W. Krabs: Stability and Controllability in Non-Autonomous
    Time-Discrete Dynamical Systems.
    Journal of Difference Equations and Applications 8, no. 12, 1107-1118 (2002).

16. W. Krabs: A Cooperative Treatment of an n-Person Cost-Goal-Game.
    Mathematical Methods of Operations Research *57*, 309-319 (2003).
17. W. Krabs, St. W. Pickl, and J. Scheffran: Optimization of an n-Person Game
    Under Linear Side Conditions.
    In: Optimization, Dynamics and Economic Analysis,
    edited by E. J. Dockner, R. F. Hartl, M. Luptacik, and G. Sorger.
    Physica-Verlag: Heidelberg, New York 2000, pp. 76 - 85.
18. U. Krause und T. Nesemann: Differenzengleichungen und
    diskrete dynamische Systeme.
    B. G. Teubner: Stuttgart und Leipzig 1999.
19. V. Lakshmikantham and D. Trigiante: Theory of Difference Equations:
    Numerical Methods and Applications.
    Academic Press, Inc.: Boston et cet. 1988.
20. J. P. La Salle: The Stability and Control of Discrete Processes.
    Springer-Verlag: New York - Berlin - Heidelberg - London - Paris - Tokio 1986.
21. J. Li: Das dynamische Verhalten diskreter Evolutionsspiele.
    Shaker Verlag: Aachen 1999.
22. G. Luenberger: Introduction to Dynamic Systems.
    John Wiley and Sons: Chichester, New York, Brisbane, Toronto 1979.
23. G. Luenberger: Introduction to Linear and Nonlinear Programming.
    Addison Wesley 1980.
24. G. Owen: On the Core of Linear Production Games.
    Mathematical Programming 9 (1975), 358 - 370.
25. St. W. Pickl: Der τ-value als Kontrollparameter. Modellierung und Analyse
    eines Joint-Implementation Programmes mithilfe der dynamischen
    kooperativen Spieltheorie und der diskreten Optimierung.
    Shaker Verlag: Aachen 1999.
26. C. Roos, T. Terlaky, J.-Ph. Vial: Theory and Algorithms for
    Linear Optimization.
    John Wiley & Sons, Chichester, 1997.
27. A. Schrijver: Theory of Linear and Integer Programming.
    John Wiley, New York 1986.
28. P. Spelucci: Numerische Verfahren der nichtlinearen Optimierung.
    Birkäuser Verlag, Boston 1993.
29. S.H. Tijs, T.S.H. Driessen: Game Theory and Cost Allocation Methods.
    Management Science, 32, no.8, 1015-1028, 1986.

Index

About the Authors

Prof. Dr. rer.nat. Werner Krabs
born 1934 in Hamburg-Altona, 1954-1959 study of mathematics, physics and astronomy at the University of Hamburg, diploma in mathematics, 1963 phd-thesis. 1967/68 visiting assistant professor at the University of Washington in Seattle. 1968 habilitation in applied mathematics at the University of Hamburg. 1970-72 professor at the RWTH Aachen. 1971 visiting associate professor at the Michigan State University in East Lansing. 1972 professor at the TH Darmstadt. 1977 visiting full professor at the Oregon State University in Corvallis. 1979-81 vice-president of the TH Darmstadt. 1986-87 chairman of the Society for Mathematics, Economy and Operations Research.

Dr. rer. nat. Stefan Wolfgang Pickl
born 1967 in Darmstadt. 1987-1993 study of mathematics, electrical engineering and philosophy at the TH Darmstadt. 1993 ERASMUS grant and diploma thesis at the EPFL Lausanne, diploma in theoretical electrical engineering. 1998 phd-thesis in mathematics at the TU Darmstadt. Dissertation award 2000 of the German Society of Operations Research. Since 2000 assistant professor at the University of Cologne.

Lecture Notes in Economics and Mathematical Systems

For information about Vols. 1–434
please contact your bookseller or Springer-Verlag

Vol. 479: L. v. Thadden, Money, Inflation, and Capital Formation. IX, 192 pages. 1999.

Vol. 480: M. Grazia Speranza, P. Stähly (Eds.), New Trends in Distribution Logistics. X, 336 pages. 1999.

Vol. 481: V. H. Nguyen, J. J. Strodiot, P. Tossings (Eds.). Optimation. IX, 498 pages. 2000.

Vol. 482: W. B. Zhang, A Theory of International Trade. XI, 192 pages. 2000.

Vol. 483: M. Königstein, Equity, Efficiency and Evolutionary Stability in Bargaining Games with Joint Production. XII, 197 pages. 2000.

Vol. 484: D. D. Gatti, M. Gallegati, A. Kirman, Interaction and Market Structure. VI, 298 pages. 2000.

Vol. 485: A. Garnaev, Search Games and Other Applications of Game Theory. VIII, 145 pages. 2000.

Vol. 486: M. Neugart, Nonlinear Labor Market Dynamics. X, 175 pages. 2000.

Vol. 487: Y. Y. Haimes, R. E. Steuer (Eds.), Research and Practice in Multiple Criteria Decision Making. XVII, 553 pages. 2000.

Vol. 488: B. Schmolck, Ommitted Variable Tests and Dynamic Specification. X, 144 pages. 2000.

Vol. 489: T. Steger, Transitional Dynamics and Economic Growth in Developing Countries. VIII, 151 pages. 2000.

Vol. 490: S. Minner, Strategic Safety Stocks in Supply Chains. XI, 214 pages. 2000.

Vol. 491: M. Ehrgott, Multicriteria Optimization. VIII, 242 pages. 2000.

Vol. 492: T. Phan Huy, Constraint Propagation in Flexible Manufacturing. IX, 258 pages. 2000.

Vol. 493: J. Zhu, Modular Pricing of Options. X, 170 pages. 2000.

Vol. 494: D. Franzen, Design of Master Agreements for OTC Derivatives. VIII, 175 pages. 2001.

Vol. 495: I Konnov, Combined Relaxation Methods for Variational Inequalities. XI, 181 pages. 2001.

Vol. 496: P. Weiß, Unemployment in Open Economies. XII, 226 pages. 2001.

Vol. 497: J. Inkmann, Conditional Moment Estimation of Nonlinear Equation Systems. VIII, 214 pages. 2001.

Vol. 498: M. Reutter, A Macroeconomic Model of West German Unemployment. X, 125 pages. 2001.

Vol. 499: A. Casajus, Focal Points in Framed Games. XI, 131 pages. 2001.

Vol. 500: F. Nardini, Technical Progress and Economic Growth. XVII, 191 pages. 2001.

Vol. 501: M. Fleischmann, Quantitative Models for Reverse Logistics. XI, 181 pages. 2001.

Vol. 502: N. Hadjisavvas, J. E. Martínez-Legaz, J.-P. Penot (Eds.), Generalized Convexity and Generalized Monotonicity. IX, 410 pages. 2001.

Vol. 503: A. Kirman, J.-B. Zimmermann (Eds.), Economics with Heterogenous Interacting Agents. VII, 343 pages. 2001.

Vol. 504: P.-Y. Moix (Ed.),The Measurement of Market Risk. XI, 272 pages. 2001.

Vol. 505: S. Voß, J. R. Daduna (Eds.), Computer-Aided Scheduling of Public Transport. XI, 466 pages. 2001.

Vol. 506: B. P. Kellerhals, Financial Pricing Models in Continuous Time and Kalman Filtering. XIV, 247 pages. 2001.

Vol. 507: M. Koksalan, S. Zionts, Multiple Criteria Decision Making in the New Millenium. XII, 481 pages. 2001.

Vol. 508: K. Neumann, C. Schwindt, J. Zimmermann, Project Scheduling with Time Windows and Scarce Resources. XI, 335 pages. 2002.

Vol. 509: D. Hornung, Investment, R&D, and Long-Run Growth. XVI, 194 pages. 2002.

Vol. 510: A. S. Tangian, Constructing and Applying Objective Functions. XII, 582 pages. 2002.

Vol. 511: M. Külpmann, Stock Market Overreaction and Fundamental Valuation. IX, 198 pages. 2002.

Vol. 512: W.-B. Zhang, An Economic Theory of Cities.XI, 220 pages. 2002.

Vol. 513: K. Marti, Stochastic Optimization Techniques. VIII, 364 pages. 2002.

Vol. 514: S. Wang, Y. Xia, Portfolio and Asset Pricing. XII, 200 pages. 2002.

Vol. 515: G. Heisig, Planning Stability in Material Requirements Planning System. XII, 264 pages. 2002.

Vol. 516: B. Schmid, Pricing Credit Linked Financial Instruments. X, 246 pages. 2002.

Vol. 517: H. I. Meinhardt, Cooperative Decision Making in Common Pool Situations. VIII, 205 pages. 2002.

Vol. 518: S. Napel, Bilateral Bargaining. VIII, 188 pages. 2002.

Vol. 519: A. Klose, G. Speranza, L. N. Van Wassenhove (Eds.), Quantitative Approaches to Distribution Logistics and Supply Chain Management. XIII, 421 pages. 2002.

Vol. 520: B. Glaser, Efficiency versus Sustainability in Dynamic Decision Making. IX, 252 pages. 2002.

Vol. 521: R. Cowan, N. Jonard (Eds.), Heterogenous Agents, Interactions and Economic Performance. XIV, 339 pages. 2003.

Vol. 522: C. Neff, Corporate Finance, Innovation, and Strategic Competition. IX, 218 pages. 2003.

Vol. 523: W.-B. Zhang, A Theory of Interregional Dynamics. XI, 231 pages. 2003.

Vol. 524: M. Frölich, Programme Evaluation and Treatment Choise. VIII, 191 pages. 2003.

Vol. 525:S. Spinler, Capacity Reservation for Capital-Intensive Technologies. XVI, 139 pages. 2003.

Vol. 526: C. F. Daganzo, A Theory of Supply Chains. VIII, 123 pages. 2003.

Vol. 527: C. E. Metz, Information Dissemination in Currency Crises. XI, 231 pages. 2003.

Vol. 528: R. Stolletz, Performance Analysis and Optimization of Inbound Call Centers. X, 219 pages. 2003.

Vol. 529: W. Krabs, S. W. Pickl, Analysis, Controllability and Optimization of Time-Discrete Systems and Dynamical Games. XII, 187 pages. 2003.

Vol. 530: R. Wapler, Unemployment, Market Structure and Growth. XXVII, 207 pages. 2003.